REALIZING THE PROMISE AND MINIMIZING THE PERILS OF AI FOR SCIENCE AND THE SCIENTIFIC COMMUNITY

Edited by

Kathleen Hall Jamieson,
William Kearney,
and
Anne-Marie Mazza

PENN

UNIVERSITY OF PENNSYLVANIA PRESS

PHILADELPHIA

Published by
University of Pennsylvania Press
Philadelphia, Pennsylvania 19104–4112
www.pennpress.org

Printed in the United States of America on acid-free
paper

10 9 8 7 6 5 4 3 2 1
Paperback ISBN: 978-1-5128-2748-4
eBook ISBN: 978-1-5128-2749-1
OA electronic ISBN: 978-1-5128-2747-7

A Cataloging-in-Publication record is available from the
Library of Congress

Realizing the Promise and Minimizing the Perils of AI for Science and the Scientific Community

CONTENTS

1. Overview and Context　　　　　　　　　　　　　　　　　1
　　Kathleen Hall Jamieson, Anne-Marie Mazza,
　　and William Kearney

2. The Value and Limits of Statements from the Scientific
Community: Human Genome Editing as a Case Study　　15
　　David Baltimore and Robin Lovell-Badge

3. Science in the Context of AI　　　　　　　　　　　　21
　　Jeannette M. Wing

4. We've Been Here Before: Historical Precedents for
Managing Artificial Intelligence　　　　　　　　　　35
　　Marc Aidinoff and David I. Kaiser

5. Navigating AI Governance as a Normative Field:
Norms, Patterns, and Dynamics　　　　　　　　　　57
　　Urs Gasser

6. Challenges to Evaluating Emerging Technologies
and the Need for a Justice-Led Approach to Shaping Innovation　99
　　Alex John London

7. Bringing Power In: Rethinking Equity Solutions for AI　127
　　Shobita Parthasarathy and Jared Katzman

8. Scientific Progress in Artificial Intelligence: History,
Status, and Futures　　　　　　　　　　　　　　147
　　Eric Horvitz and Tom M. Mitchell

9. Perspectives on AI from Across the Disciplines   195

     David Baltimore, Vinton G. Cerf, Joseph S. Francisco, Barbara J. Grosz,
     John L. Hennessy, Eric Horvitz, Kathleen Hall Jamieson, Marcia K.
     McNutt, Saul Perlmutter, William H. Press, Jeannette M. Wing,
     and Michael Witherell

10. Protecting Scientific Integrity in an Age of Generative AI   221

     Wolfgang Blau, Vinton G. Cerf, Juan Enriquez, Joseph S. Francisco,
     Urs Gasser, Mary L. Gray, Mark Greaves, Barbara J. Grosz,
     Kathleen Hall Jamieson, Gerald H. Haug, John L. Hennessy,
     Eric Horvitz, David I. Kaiser, Alex John London, Robin Lovell-Badge,
     Marcia K. McNutt, Martha Minow, Tom M. Mitchell, Susan Ness,
     Shobita Parthasarathy, Saul Perlmutter, William H. Press,
     Jeannette M. Wing, and Michael Witherell

11. Safeguarding the Norms and Values of Science in the
Age of Generative AI   229

     Kathleen Hall Jamieson and Marcia K. McNutt

Appendix 1. List of Retreatants   253

Appendix 2. Biographies of Framework Authors, Paper
Authors, and Editors   257

Index   267

Realizing the Promise and Minimizing
the Perils of AI for Science and the
Scientific Community

CHAPTER 1

Overview and Context

Kathleen Hall Jamieson, Anne-Marie Mazza, and William Kearney

Prior to the advent of ChatGPT, there were several moments when artificial intelligence (AI) captured news headlines. One such instance occurred in 1997 when IBM's Big Blue computer won a chess match against the reigning grandmaster of the game. Then, in 2011, IBM's Watson beat "the best human Jeopardy! player ever." Six years later AlphaGo defeated the world's top player of one of the most complicated board games, Go. A *New York Times* article on AlphaGo's victory began with the line, "It isn't looking good for humanity."[1]

The Times's lead sentence about AI outsmarting humans portended the worries that would emerge when the world awoke to the power, promise, and peril of artificial intelligence. That awakening occurred in late 2022 when OpenAI released ChatGPT, an AI chatbot capable of generating conversational answers and analyses, as well as images, in response to user questions and prompts. This generative AI is built with computational procedures, including large language models, that train on vast bodies of human-created and curated data, including huge amounts of scientific literature. It also has the ability to generate novel syntheses and ideas of its own that "push the expected boundaries of automated content creation."[2]

Generative AI is accelerating breakthrough progress in science, perhaps best highlighted by Deep Mind's AlphaFold, an AI tool that accurately predicts the unique structure of proteins, a process that in the past took many years and hundreds of thousands of dollars to accomplish. At the same time, generative AI is raising concerns about how its use in research may under-mine core norms and values of science, including accountability, transparency, replicability, and human responsibility. In addition, generative AI is still plagued on occasion by nonsensical or inaccurate output, known as hallucinations. There also is a risk that the output can be biased and could reinforce long-standing injustices, inequalities, and inequities in society. Generative AI may also be used to further the proliferation of misinformation and disinformation.

To remark, as technology experts have, that AI is "evolving at a very rapid pace," is an understatement; in fact, as this book was about to go to press, a new version of ChatGPT was said to be able to "reason" through science, math, and coding challenges.[3] The sudden advances in artificial intelligence, and generative AI in particular, with new versions of chatbots and other AI tools being unveiled every few months, are putting increased pressure on the scientific community and policymakers to monitor the advances and consider their implications for research and society at large, and not just in the short term. As a 2022 report, Fostering Responsible Computing Research, from the National Academies of Sciences, Engineering, and Medicine, Fostering Responsible Computing Research, reminds us, "The concerns at the beginning of a technology's developmental lifecycle are not the same as the ones that surface after wide-scale deployment."[4]

Responding to the rapid development and deployment of artificial intelligence and generative AI models and the growing need for thoughtful consideration of their implications for the scientific community, in Summer 2023, National Academy of Sciences (NAS) President Marcia McNutt, Annenberg Public Policy Center (APPC) Director and Sunnylands Program Director Kathleen Hall Jamieson, and Sunnylands President David Lane invited just over two dozen experts to a two-day virtual retreat (November 29–30, 2023) followed by an in-person one at Sunnylands in Rancho Mirage, California (February 8–10, 2024), to consider governance of AI and its rapid diffusion throughout society and, in particular, across the scientific research enterprise.[5] Background papers—which form the core of this book—on topics such as the evolution and current governance of AI, how the scientific community responded to past technological breakthroughs, and the societal implications, including effects on equity, of AI and other emerging technologies, were commissioned to inform these deliberations.[6] (A list of participants in one or both of the convenings, including the authors of the background papers included in this volume, can be found in Appendix 1).

Since 2015, the Annenberg Foundation Trust at Sunnylands, the APPC, and the NAS have partnered to fulfill Sunnylands's mission to host "meetings of leaders and specialists in the major medical and scientific associations and institutions for the purpose of promoting and facilitating the exchange of ideas . . . to make advancements . . . for the common good and the public interest."[7] Joined occasionally by the National Academy of Medicine, these partners have convened retreats at which leaders in science, academia, business, medical ethics, the judiciary and the bar, government, and scientific publishing identified ways to protect the integrity of science;[8] increase the transparency of authors' contributions to scholarly publications;[9] articulate the principles that should guide scientific practice to ensure that science works at the frontiers of human knowledge in an ethical way; and protect the courts from inadvertent as well as deliberate misstatements about scientific knowledge. Plans for creation of the National Academy of Sciences' Strategic Council for Research Excellence, Integrity, and Trust were birthed at an NAS-APPC-Sunnylands retreat,[10] as were recommendations to protect the integrity of survey research.[11]

In the past, as society grappled with the implications of technologies ranging from nuclear energy to recombinant DNA, CRISPR-Cas 9 gene editing, dual use research of concern, and neural organoids and chimeras, the scientific community often developed practices designed to increase adherence to the norms that have protected the integrity of each new form of scientific exploration, development, and deployment. In the process, scientists expanded their community's repertoire of mechanisms designed to advance emerging science and technology while safeguarding the integrity of science and the well-being of the nation and its people.

Leading to the development of an NIH-approved biosafety framework, the 1975 Asilomar Conference on Recombinant DNA confirmed the importance of transparency and self-regulation among scientists involved in gene-splicing technology. The *Belmont Report* (1979), developed by the National Commission for the Protection of Human Subjects of Biomedical and Behavioral Research, set respect for persons, beneficence, and justice as core ethical principles for scientists involved in human subjects research in biomedicine and led to the establishment at research institutions of Institutional Review Board (IRB) processes and checks and balances based on those principles. In the process, it added the concept of informed consent and assessment of risks and benefits to the vocabulary of researchers.

Such past efforts remind us, as do the essays in this volume, that even as our understandings of emerging technologies and of their implications evolve, science's commitment to core norms and values must remain steadfast. These reports also remind us that ethical, equitable, accountable, transparent science is the by-product of a vigilant scientific community that proactively engages the public.

Tasked with both exploring emerging challenges posed by the use of AI in research and charting a path forward for the scientific community, participants in the AI retreats included experts in behavioral and social sciences, ethics, biology, physics, chemistry, mathematics, and computer science, as well as leaders in higher education, law, governance, and science publishing and communication. Included in their ranks were three Nobel laureates and fourteen members of the National Academy of Sciences, the National Academy of Engineering, or National Academy of Medicine.[12]

In fashioning their work, the NAS-APPC-Sunnylands retreatants drew on the lessons learned from earlier workshops, reports, and consensus statements from the National Academies of Science, Engineering, and Medicine, including Fostering Integrity in Research (2017),[13] Reproducibility and Replicability in Science (2019),[14] Fostering Responsible Computing Research: Foundations and Practices (2022),[15] Automated Research Workflows for Accelerated Discovery (2022),[16] a National Academies AI for Scientific Discovery Workshop (October 12–13, 2023),[17] and National Academy of Medicine's Generative AI and LLMs in Health and Medicine Workshop (October 25, 2023).[18]

The retreatants' deliberations were informed as well by the commissioned background papers and by presentations from Nobel Laureates Harold Varmus, Lewis Thomas University Professor of Medicine at Weill Cornell Medical College, and David Baltimore, Distinguished Professor of Biology at Caltech, about efforts by the scientific community to deal with the challenges posed by potential pandemic pathogens and emergent technologies such as human genome editing. Additionally, Baltimore and Robin

Lovell-Badge, head of the laboratory of stem cell biology and development genetics at the Francis Crick Institute in London, discussed the processes that led to the three International Summits on Human Genome Editing. Those convenings created consensus statements establishing processes and ethical principles to guide research and the use of human genome editing techniques, engage the public, and protect future generations against negative consequences. A digest of insights from Baltimore and Lovell-Badge forms Chapter 2 of the book. Chapter 11, "Safeguarding the Norms and Values of Science in the Age of Generative AI," by conveners Kathleen Hall Jamieson and Marcia McNutt explores the guiding norms and values of science at issue in the working group's call for the scientific community to protect scientific integrity in the age of generative AI by remaining "steadfast in honoring the guiding norms and values of science."

The letter inviting participants to the two-stage retreats provisionally adopted the Association for the Advancement of Artificial Intelligence (AAAI) definition of artificial intelligence as "the mechanisms underlying thought and intelligent behavior and their embodiment in machines." The invitational letter also forecast that the retreatants' deliberations would "build from and contribute to the revision of draft commissioned papers that will provide: 1) a historical perspective on how society has prepared and managed emerging transformative technologies; 2) philosophical/ethical lenses used to analyze and evaluate emerging technologies; 3) an overview of recently proposed AI frameworks, laws, principles, and guidelines; 4) equity and inclusion issues associated with AI; 5) an assessment of the current state of scientific/technical advances in AI, hurdles and potential, and concerns its capacities raise; and 6) challenges and opportunities associated with creating and analyzing large data sets." (Brief biographical statements on authors whose work is included in this book can be found in Appendix 2. See also Figure 1.1.)

Expanding on the AAAI definition, the retreatants presupposed with Eric Horvitz, Chief Scientific Officer of Microsoft, and Tom Mitchell, Founders University Professor at Carnegie Mellon University (see Chapter 8) that "Artificial Intelligence (AI) refers to a field of endeavor as well as a constellation of technologies," a notion consistent with the one set forth in 15 U.S.C. 9401(3).[19]

Figure 1.1. The Annenberg Foundation Trust at Sunnylands in Partnership with the Annenberg Public Policy Center University of Pennsylvania. Standing (L–R): Shobita Parthasarathy, Distinguished Professor of Public Policy, Univ. of Michigan; David J. Lane, President, The Annenberg Foundation Trust at Sunnylands; Michael Witherell, Director, Lawrence Berkeley National Laboratory; Tom Mitchell, Distinguished Fellow, Carnegie Mellon; Mark Greaves, Executive Director, AI2050; David Kaiser, Distinguished Professor of History, MIT; William Kearney, Executive Director of News, NASEM; Anne-Marie Mazza, Senior Director, NASEM; Juan Enriquez, Managing Director, Excel Venture Management; Eric Horvitz, Chief Scientific Officer, Microsoft; William Press, Distinguished Professor of Computer Science, UT Austin; Saul Perlmutter, Distinguished Professor of Physics, UC Berkeley; Urs Gasser, Professor of Public Policy, TUM; Kathleen Hall Jamieson, Director, Annenberg Public Policy Center; Alex John London, Distinguished Professor of Ethics, Carnegie Mellon; Kathleen Doherty, Chief Strategy and Retreats Office, The Annenberg Foundation Trust at Sunnylands; Jeannette M. Wing, Executive Vice President for Research, Columbia. Seated (L–R): Martha Minow, Distinguished Professor of Law, Harvard; Susan Ness, Former Commissioner, FCC; Joseph Francisco, Distinguished Professor of Earth and Environmental Science, UPenn; Marcia McNutt, President, NAS; John Hennessy, President Emeritus, Stanford; Barbara Grosz, Distinguished Professor Natural Sciences, Harvard; Gerald Haug, President, Leopoldina; Mary L. Gray, Senior Principal Research, Microsoft Research.

With this in mind, the Sunnylands Statement (see Chapter 10) that emerged from the AI retreats built upon the understanding that "generative AI systems are constructed with computational procedures that learn from large bodies of human-authored and curated text, imagery, and analyses, including expansive collections of scientific literature. The systems are used to perform multiple operations, such as problem-solving, data analysis, interpretation of textual and visual content, and the generation of text, images, and other forms of data. In response to prompts and other directives, the systems can provide users with coherent text, compelling imagery, and analyses, while also possessing the capability to generate novel syntheses and ideas that push the expected boundaries of automated content creation."

As a means of "understanding the opportunities and risks associated with AI today," in Chapter 4, "We've Been Here Before: Historical Precedents for Managing Artificial Intelligence," Marc Aidinoff, Research Associate at the Institute for Advanced Learning, and David Kaiser, Germeshausen Professor of the History of Science at the Massachusetts Institute for Technology, consider the ways in which the scientific community dealt with three historical episodes: "the early nuclear-weapons complex during the 1940s and 1950s; biotechnology, biomedicine, and the implementation of various safeguards in the 1970s; and the adoption and oversight of forensic technologies within the US legal and criminal-justice systems over the course of the past century." In their digest in *Issues in Science and Technology*, they argue that "artificial intelligence needs ongoing and meaningful democratic oversight" which can be informed by understanding these historical episodes.[20]

In Chapter 5, "Navigating AI Governance as a Normative Field: Norms, Patterns, and Dynamics," Urs Gasser, Professor of Public Policy, Governance, and Innovative Technology at the Technical University of Munich, addresses the "rapidly evolving and complex ecosystem" that surrounds AI and identifies the a variety of tools available to decision-makers as they "seek to anticipate, analyze, and address harms and risks associated with the accelerating pace of AI development, deployment, and use while harnessing its potential for human, society, and the planet at large." This includes both ethical and technical standards. In his digest in *Issues in Science and Technology*, Gasser calls for AI governance that "leaves space for development and learning," prioritizes interoperability, and invests in implementation capacity.[21]

Alex John London, K&L Gates Professor of Ethics and Computational Technologies at Carnegie Mellon University, then makes the case for a justice-led framework when evaluating innovations such as generative AI in Chapter 6, "Challenges to Evaluating Emerging Technologies and the Need for a Justice-Led Approach to Shaping Innovation." A justice-led focus, he argues, is better able to identify and evaluate "(a) quintessentially social or higher-order effects (such as network-level or institutional level effects), (b) the role of a larger number of stakeholders who shape the innovation ecosystem in more indirect ways, and (c) some of the positive ethical claims of individuals that are relevant to evaluating innovation." In his digest in *Issues in Science and Technology*, London argues that a justice-led framework will promote "social arrangements that better secure people's freedom in the face of technological change."[22]

In Chapter 7, "Bringing Power In: Rethinking Equity Solutions for AI," Shobita Parthasarathy, Professor of Public Policy and Women's and Gender Studies at the University of Michigan, and Jared Katzman, PhD student at the University of Michigan School of Information, draw our attention to growing concerns that AI is "exacerbating social inequity and injustice." Their essay explores the responses of "policymakers, academics, and the technical community," including the Blueprint for an AI Bill of Rights proposed by the Biden administration. That document "recommends identifying statistical biases in datasets, designing systems to be more transparent and explainable in their decision-making, incorporating proactive equity assessments into system design, including input from diverse viewpoints and identities, ensuring accessibility for people with disabilities, pre-deployment and ongoing disparity testing and mitigation, and clear oversight."[23] They argue that many of such initiatives fall short because they fail to address "social inequalities that shape the landscape of technology development, use, and governance, including the concentration of economic and political power in a handful of technology companies and the systematic devaluation of lay contributions and perspectives, especially from those who have been historically marginalized." Instead, as they argue in *Issues in Science and Technology*, AI regulators ought to "seek out partnerships with marginalized communities" in order to understand "power imbalances at the root of concerns surrounding AI bias and discrimination."

Horvitz and Mitchell synthesize the journey of AI's decades of "innovation with empirical studies and prototypes, the development of theoretical principles, and shifts among paradigms" in Chapter 8, "Scientific Progress in Artificial Intelligence: History, Status, and Futures." In the process, they provide a lens on understanding "the technical evolution of different approaches to representing and reasoning with data and knowledge," the machine learning foundations of today's AI, as well as of discriminative and generative models, supervised, unsupervised, and self-supervised learning, and the inflection point for AI occasioned by deep learning. They also define key concepts and research directions before looking to a second inflection point: generative AI and charting its research, directions, trends, and key opportunities with applications for discriminative and generative AI.

As a complement to these efforts, members of the AI working group, Michael Witherell, Director of the Lawrence Berkeley National Laboratory, and William Press, the Leslie Surginer Professor of Computer Science and Integrative Biology at the University of Texas at Austin, planned an April 27 symposium for the 2024 annual meeting of the National Academy of Sciences moderated by working group member Jeannette Wing, Executive Vice President for Research and Professor of Computer Science at Columbia University, whose presentation in the symposium is the basis for Chapter 3, "Science in the Context of AI" (Figures 1.2 and 1.3).

"Much of the conversation we hear today about Artificial Intelligence (AI) focuses on fears concerning loss of privacy, lack of transparency and accountability, increase in inequality, and other social and economic issues," noted the symposium planners William Press and Michael Witherell. "The widespread availability of generative AI is fueling much of this debate. However, AI is more than just large language models, and in fact versions of AI have been fueling scientific discovery and exploration for several decades now."[24]

Titled "AI and Scientific Discovery," the symposium provided "an opportunity to hear from speakers at the forefront of developing AI to advance research by automating workflows, finding patterns in large and complex data sets, mitigating human bias, improving models, speeding up tedious tasks, and exploring domains inhospitable to humans."

Figure 1.2. Marcia McNutt, President of the National Academy of Sciences, discusses the retreat on AI and science.

Figure 1.3. Standing: Jeannette M. Wing, Executive Vice President for Research and Professor at Columbia University, moderated a symposium titled "AI and Science" at the 161st Annual Meeting of the National Academy of Sciences on April 27, 2024. Seated (L–R): Jennifer Listgarten, UC Berkeley: "The Perpetual Motion Machine of AI-Generated Data and the Distraction of ChatGPT as a 'Scientist'"; Michael Pritchard, UC Irvine and Nvidia: "The Impact on Weather Prediction and Climate Simulation"; Daphne Koller, Insitro: "Using AI to Accelerate Drug Discovery"; Pushmeet Kohli, Google DeepMind: "A New Era of Digital Biology."

Joining Wing in exploring both the promise of and various possible futures for AI-assisted research were four panelists:

- Pushmeet Kohli, Vice President of Research at Google DeepMind
- Daphne Koller, Founder and CEO of Insitro
- Michael Pritchard, Director of Climate Simulation Research at NVIDIA and Professor at the University of California, Irvine
- Jennifer Listgarten, Professor of Computer Science at the University of California, Berkeley

To provide a snapshot of the ways in which AI was affecting science, at both the virtual and in-person retreats, National Academy of Sciences, National Academy of Engineering, and National Academy of Medicine members of the Sunnylands working group shared their thoughts on the ways in which AI was affecting or might affect their work. As a means of preserving a sense of the ways in which AI was transforming scientific research in the months in which the retreatants were fashioning the calls for action found in their PNAS editorial, we include a digest of their thoughts in Chapter 9, "Perspectives on AI from Across the Disciplines."

The working group's editorial statement "Protecting Scientific Integrity in an Age of Generative AI" was published in the Proceedings of the National Academy of Sciences (PNAS) on May 21, 2024 and is included as Chapter 10 in this volume.[25] The editorial emphasizes that advances in generative AI represent a transformative moment for science—one that will accelerate scientific discovery but also challenge core norms and values of science, such as accountability, transparency, replicability, and human responsibility. "We welcome the advances that AI is driving across scientific disciplines, but we also need to be vigilant about upholding long-held scientific norms and values," said National Academy of Sciences President Marcia McNutt, one of the coauthors of the editorial. "We hope our paper will prompt reflection among researchers and set the stage for concerted efforts to protect the integrity of science as generative AI increasingly is used in the course of research."[26]

Urging the scientific community to follow five principles of human accountability and responsibility when using artificial intelligence in research, the editorial advocated: transparent disclosure and attribution; verification of AI-generated content and analyses; documentation of AI-generated data; a focus on ethics and equity; and continuous monitoring, oversight, and public engagement.

Its twenty-four authors also called on the National Academy of Sciences to establish a Strategic Council on the Responsible Use of AI in Science to provide ongoing guidance and oversight on responsibilities and best practices as the technology evolves. The proposed strategic council should be established by the National Academies of Sciences, Engineering, and Medicine, the authors recommended, and should coordinate with the scientific community and provide regularly updated guidance on the appropriate uses of AI. The council should study, monitor, and address the evolving use of AI in science; new ethical and societal concerns, including equity; and emerging threats to scientific norms. It should also share its insights across disciplines and develop and refine best practices.

This edited volume capsulizes the discussions that shaped the statement "Protecting Scientific Integrity in an Age of Generative AI" and provides a snapshot both of the state of AI science in Spring 2024 and of the efforts by leaders of the scientific community to ensure that the use of AI in research is pursued in a responsible manner. We hope it will provide a foundation for consideration of this fast moving and transformative technology.

Notes

1. Paul Mozur, "Google's AlphaGo Defeats Chinese Go Master in Win for A.I.," *New York Times*, May 23, 2017, https://www.nytimes.com/2017/05/23/business/google-deepmind -alphago-go-champion-defeat.html.

2. Mozur, "Google's AlphaGo Defeats Chinese Go Master."

3. OpenAI, "Introducing OpenAI o1," September 12, 2024, https://openai.com/o1/.

4. *Fostering Responsible Computing Research*, National Academies Press eBooks (2022), 41, https://doi.org/10.17226/26507.

5. Partial support for the Sunnylands Retreats was provided by the Ralph J. Cicerone and Carol M. Cicerone Endowment for NAS Missions.

6. Chapter 4: "We've Been Here Before: Historical Precedents for Managing Artificial Intelligence," by Marc Aidinoff and David Kaiser; Chapter 5: "Navigating AI Governance as a Normative Field: Norms, Patterns, and Dynamics," by Urs Gasser; Chapter 6: "Challenges to Evaluating Emerging Technologies and the Need for a Justice-Led Approach to Shaping Innovation," by Alex John London; Chapter 7: "Bringing Power In: Rethinking Equity Solutions for AI," by Shobita Parthasarathy and Jared Katzman; Chapter 8: "Scientific Progress in Artificial Intelligence: History, Status, and Futures," by Eric Horvitz and Tom Mitchell.

7. Annenberg Public Policy Center, "National Academies, Sunnylands, and APPC Host Retreats on Protecting the Integrity of Science," August 7, 2020, https://www.annenberg publicpolicycenter.org/nas-appc-sunnylands-retreats-integrity-science/.

8. Bruce Alberts et al., "Self-Correction in Science at Work," *Science* 348, no. 6242 (June 26, 2015): 1420–1422, https://doi.org/10.1126/science.aab3847.

9. Marcia K. McNutt et al., "Transparency in Authors' Contributions and Responsibilities to Promote Integrity in Scientific Publication," *Proceedings of the National Academy of Sciences* 115, no. 11 (February 27, 2018): 2557–2560, https://doi.org/10.1073/pnas.1715374115.

10. National Academies of Science, Engineering, and Medicine, "New Strategic Council for Research Excellence, Integrity, and Trust Established by National Academy of Sciences to Support the Health of the Research Enterprise," National Academies, July 13, 2021, https:// www.nationalacademies.org/news/2021/07/new-strategic-council-for-research-excellence -integrity-and-trust-established-by-national-academy-of-sciences-to-support-the-health -of-the-research-enterprise.

11. Kathleen Hall Jamieson et al., "Protecting the Integrity of Survey Research," *PNAS Nexus* 2, no. 3 (March 1, 2023), https://doi.org/10.1093/pnasnexus/pgad049.

12. Some individuals represent several of the Academies.

13. *Fostering Integrity in Research*, National Academies Press eBooks (2017), https://doi .org/10.17226/21896.

14. *Reproducibility and Replicability in Science*, National Academies Press eBooks (2019), https://doi.org/10.17226/25303.

15. *Fostering Responsible Computing Research*.

16. *Automated Research Workflows for Accelerated Discovery*, National Academies Press eBooks (2022), https://doi.org/10.17226/26532.

17. *AI For Scientific Discovery*, National Academies Press eBooks (2024), https://doi.org /10.17226/27457.

18. National Academies of Medicine, "Generative AI & LLMs in Health & Medicine," October 25, 2023, https://nam.edu/event/generative-ai-llms-in-health-medicine/.

19. This definition of "artificial intelligence" or "AI" is set forth in 15 U.S.C. 9401(3). "A machine-based system that can, for a given set of human-defined objectives, make predictions, recommendations, or decisions influencing real or virtual environments. Artificial intelligence systems use machine- and human-based inputs to perceive real and virtual environments; abstract such perceptions into models through analysis in an automated manner; and use model inference to formulate options for information or action. See *Managing Misuse Risk for Dual-Use3 Foundation Models*, U.S. AI Safety Institute, https://doi.org/10 .6028/NIST.AI.800-1.ipd.

20. Digests of the commissioned papers were published in *Issues in Science and Technology*. See Marc Aidinoff and David Kaiser, "Novel Technologies and the Choices We Make: Historical Precedents for Managing Artificial Intelligence," *Issues in Science and Technology*, May 21, 2024, https://doi.org/10.58875/buxb2813; Urs Gasser, "Governing AI with Intelligence," *Issues in Science and Technology*, May 21, 2024, https://doi.org/10.58875/ awjg1236; Alex John London, "A Justice-Led Approach to AI Innovation," *Issues in Science and Technology*, May 21, 2024, https://doi.org/10.58875/knrz2697; Shobita Parthasarathy and Jared Katzman, "Bringing Communities in, Achieving AI for All," *Issues in Science and Technology*, May 21, 2024, https://doi.org/10.58875/slrg2529.

21. Gasser, "Governing AI with Intelligence."

22. London, "A Justice-Led Approach to AI Innovation."

23. Office of Science and Technology Policy, *Blueprint for an AI Bill of Rights*, The White House, October 2022, https://www.whitehouse.gov/wp-content/uploads/2022/10/Blueprint -for-an-AI-Bill-of-Rights.pdf.

24. National Academy of Science (@theNASciences), "NAS 161st Annual Meeting—Symposium AI and Scientific Discovery," YouTube video (May 24, 2024), https://www .youtube.com/watch?v=G43Em6ELaiE.

25. Wolfgang Blau et al., "Protecting Scientific Integrity in an Age of Generative AI," *Proceedings of the National Academy of Sciences* 121, no. 22 (May 21, 2024), https://doi.org/10 .1073/pnas.2407886121.

26. National Academies of Science, Engineering, and Medicine, "Human Accountability and Responsibility Needed to Protect Scientific Integrity in an Age of AI, Says New Editorial," National Academies, May 21, 2024, https://www.nationalacademies.org/news/2024 /05/human-accountability-and-responsibility-needed-to-protect-scientific-integrity-in-an -age-of-ai-says-new-editorial?mc_cid=24fa0dce18&mc_eid=e9e4cc3749.

CHAPTER 2

The Value and Limits of Statements from the Scientific Community: Human Genome Editing as a Case Study

David Baltimore and Robin Lovell-Badge

n late November 2018, when we arrived in Hong Kong for the Second International Summit on Human Genome Editing, we were met with the news that a researcher, He Jiankui of the Southern University of Science and Technology in Shenzhen, China, had edited embryos that developed into twin girls who were born just a month earlier. We had invited Dr. He to speak at the summit as part of an ongoing global discussion, started in 2015, about the appropriate use of breakthrough gene editing tools, including CRISPR-Cas9. At the time we sent our invitation to Dr. He we were not aware of his experiment. When we did become aware of this in Hong Kong, all of us on the summit organizing committee, a group convened by the US National Academy of Sciences and National Academy of Medicine, UK Royal Society, and the Academy of Sciences of Hong Kong, were deeply concerned. Fortunately, the statement issued by the organizing committee (of which we were also members) at the end of the first summit on human genome editing, held in Washington, DC, in late 2015, provided a guidepost to help us respond effectively to the news of Dr. He's experiment.

The first summit convened in 2015 by the US National Academy of Sciences and National Academy of Medicine, the UK Royal Society, and the Chinese Academy of Sciences, drew worldwide attention. After three days of lively discussions, the organizing committee issued a statement, in which the key conclusion declared:

> It would be irresponsible to proceed with any clinical use of germline editing unless and until (i) the relevant safety and efficacy issues have been resolved, based on appropriate understanding and balancing of risks, potential benefits, and alternatives, and (ii) there is broad societal consensus about the appropriateness of the proposed application. Moreover, any clinical use should proceed only under appropriate regulatory oversight. At present, these criteria have not been met for any proposed clinical use: the safety issues have not yet been adequately explored; the cases of most compelling benefit are limited; and many nations have legislative or regulatory bans on germline modification.[1]

When we learned on the eve of our 2018 summit in Hong Kong that Dr. He had used CRISPR/Cas-9 to edit the embryos of newly born twin girls, specifically targeting a gene, CCR5, that codes for a protein that HIV-1 uses to enter cells, a few members of our organizing committee met with him to ask exactly what he had done. What we learned from him indicated that he had acted irresponsibly, given that the conditions set forth in the 2015 statement had not yet been met (nor have they currently been met). We agreed, however, that in the interest of scientific openness, Dr. He should be allowed to remain on the agenda and given an opportunity to report on his work. When he spoke on the second day of the summit, there were nearly 100 reporters and photographers in the auditorium, along with approximately 400 attendees, and a worldwide audience of over one million watching a live webcast. Immediately after Dr. He's presentation, I (David Baltimore, as chair of the 2018 summit organizing committee) took the lectern to condemn He's experiment as "irresponsible," and to criticize him for his lack of transparency; then I added: "There has been a failure of self-regulation by the scientific community." Although our 2015 statement provided important principles that we expected the scientific community to follow, it had failed to stop a rogue scientist from pursuing heritable genome editing.

At the end of the Hong Kong summit, the organizing committee, comprising representatives from eight countries, issued a new statement updating the 2015 statement. While expressing deep concern with Dr. He's reported experiment, we also took note of the progress that had been made in further developing genome editing tools, the attention that various countries were giving to genome editing, and the need for ongoing public engagement. We noted the "rapid advance of somatic gene editing into clinical trials," but at the same time, emphasized that we "continue to believe that proceeding with any clinical use of germline editing remains irresponsible at this time."[2] (Somatic gene editing is the editing of genes in adult cells, whereas heritable genome editing via germline cells involves edits that may be passed onto future generations.)

The word "irresponsible" in both statements was not used lightly. Responsibility is a hallmark of scientific excellence and integrity. As science advances and our understanding increases, it is essential that the scientific community maintains high expectations for its members.

Statements such as these are not meant to be set in stone but rather should be updated as the science moves forward and societal concerns change. Our first statement, therefore, acknowledged that "as scientific knowledge advances and societal views evolve, the clinical use of germline editing should be revisited on a regular basis."[3] Thus, the 2018 statement called for a more defined approach to determine whether germline editing in clinical settings could someday be permitted in accordance with scientific and medical norms and values. We stated that

> the scientific understanding and technical requirements for clinical practice remain too uncertain and the risks too great to permit clinical trials of germline editing at this time. Progress over the last three years and the discussions at the current summit, however, suggest that it is time to define a rigorous, responsible translational pathway toward such trials.[4]

In addition, the statement emphasized that the pathway must adhere to

> widely accepted standards for clinical research, including criteria
> articulated in genome editing guidance documents published in
> the last three years. Such a pathway will require establishing
> standards for preclinical evidence and accuracy of gene modifica-
> tion, assessment of competency for practitioners of clinical trials,
> enforceable standards of professional behavior, and strong
> partnerships with patients and patient advocacy groups.[5]

Following the 2018 summit numerous organizations issued reports detailing such a pathway.[6] When we convened in London in 2023 for the Third
International Summit on Human Genome editing, chaired by Robin Lovell-
Badge, coauthor of this essay, the world had just begun emerging from a
global pandemic. The organizing committee, which included representatives from eleven countries, sought to continue the discussion about the
science and ethics of genome editing, given the tremendous advances that
had been made in somatic cell editing. We also drew even greater attention
to issues of access, cost, and equity, especially given the price of emerging
gene editing treatments.

Just as we did at the first two summits, the organizing committee issued
a statement at the conclusion of the third. That statement noted "remarkable" progress in somatic genome editing, based on demonstrations, including a first-person account by a patient, that it could be used to cure once
incurable diseases. Further, the statement indicated that to realize the full
therapeutic benefits of somatic editing, "research is needed to expand the
range of diseases it can treat, and to better understand risks and unintended
effects."[7] It noted that "the extremely high costs of current somatic gene
therapies are unsustainable" and identified an urgent need for "a global
commitment to affordable, equitable access to these treatments."[8]

With respect to heritable genome editing, the committee reiterated that
it "remains unacceptable at this time."[9] Moreover, we added that: "Public
discussions and policy debates continue and are important for resolving
whether this technology should be used. Governance frameworks and ethi-

cal principles for the *responsible* use of heritable human genome editing are not in place. Necessary safety and efficacy standards have not been met."[10]

Having set the foundation at our 2015 summit for the pursuit of somatic gene editing technologies while expressing extreme caution about the use of heritable editing, over the next eight years the statements were able to reaffirm, clarify, and take note of advances in science and societal responses. Each statement built on the knowledge that the scientific community had gained and the ever-evolving expectations and demands of society to participate in decisions about the clinical applications of the science. Even though it is difficult to obtain universal consensus on contentious issues such as this, the statements were authored by diverse organizing committees and with the input of diverse audiences. They served as a reference point for the scientific community and the public and spurred ongoing engagement with diverse stakeholders. Although we do not have the capacity for global monitoring, and despite Dr. He's demonstration of the relative ease at which heritable genome editing could be done, albeit very badly, we are unaware at this time of any additional attempts at human heritable germline editing. We are reluctant to credit the summit statements for this restraint, but we do believe that such statements provide guidance and grounding for the scientific community in its consideration of emerging technologies. Artificial intelligence, especially generative AI, is a case in point. This technology will likely have huge impacts on all aspects of science and society. More stringent governance may be necessary, but in the meantime, proposing guiding principles for AI's development and deployment in research is an appropriate step for leaders in the scientific community to take to convey that it should be utilized responsibly in ways that uphold the integrity and norms of science and bolster public trust.

Notes

1. Organizing Committee for the International Summit on Human Gene Editing, "On Human Gene Editing: International Summit Statement," National Academies, December 3, 2015, https://www.nationalacademies.org/news/2015/12/on-human-gene-editing-international-summit-statement.

2. "Statement by the Organizing Committee of the Second International Summit on Human Genome Editing," National Academies, November 28, 2018, https://www.nationalacademies.org/news/2018/11/statement-by-the-organizing-committee-of-the-second-international-summit-on-human-genome-editing.

3. "Statement by the Organizing Committee of the Second International Summit on Human Genome Editing."

4. "Statement by the Organizing Committee of the Second International Summit on Human Genome Editing."

5. "Statement by the Organizing Committee of the Second International Summit on Human Genome Editing."

6. National Academy of Medicine, National Academy of Sciences, and the Royal Society, *Heritable Human Genome Editing* (Washington, DC: National Academies Press, 2020), https://nap.nationalacademies.org/catalog/25665/heritable-human-genome-editing; WHO Expert Advisory Committee on Developing Global Standards for Governance and Oversight of Human Genome Editing, *Human Genome Editing: A Framework for Governance* (World Health Organization, 2021), https://iris.who.int/bitstream/handle/10665/342484/9789240030060-eng.pdf?sequence=1&isAllowed=y.

7. "Statement of the Organising Committee of the Third International Summit on Human Genome Editing," The Royal Society, March 8, 2023, https://royalsociety.org/news/2023/03/statement-third-international-summit-human-genome-editing/#:~:text=Heritable%20human%20genome%20editing%20remains,editing%20are%20not%20in%20pl.

8. "Statement of the Organising Committee of the Third International Summit on Human Genome Editing."

9. "Statement of the Organising Committee of the Third International Summit on Human Genome Editing."

10. "Statement of the Organising Committee of the Third International Summit on Human Genome Editing."

CHAPTER 3

Science in the Context of AI

Jeannette M. Wing

Generative AI took even the computer science community by surprise. To put this disruption in context, let's start at the beginning, surf through two waves of AI, and then situate science in this time line. With selective highlights, I offer a compressed history of AI; a simplified view of the transformer architecture, which underlies generative AI; and a bird's eye view of how AI can benefit science.

Part 1: AI Time Line

In 1950, Alan Turing, considered the father of modern computer science, proposed the Turing Test: If a human interacts with a machine, and the human cannot tell the difference between interacting with the machine and interacting with another human being, then the machine passes the Turing Test. We could thus consider this machine as exhibiting, in some sense, human intelligence.

The year 1956 marks not only when the term *artificial intelligence* was born but also the start of AI as an academic pursuit. At a summer conference held in Dartmouth, the participants asked whether we could build a machine that mimics the behavior of humans. This grand goal of AI was recognized early on as too difficult to achieve. Thus, subfields of AI splintered off, representing subtasks of human intelligence: computer vision for vision, speech recognition and natural language processing for language, and robotics for mobility and manipulation. Other subtasks such as logical reasoning and abstract reasoning found common ground with subfields of computer science, including theorem proving, formal methods, and programming languages. The first wave of AI, during the last half of the last century, is signified by representing knowledge symbolically, not numerically, and representing reasoning by rules. For example, with the following rule:

$$\text{Man}(X) \Rightarrow \text{Mortal}(X)$$

if Socrates is a man, then I can conclude Socrates is mortal.

These rule-based methods led to what are called "expert systems," which by the late 1990s found their way into scientific domains. The first expert system project, Dendral,[1] began in 1965 with the goal of capturing knowledge about organic chemistry so chemists could identify unknown organic molecules "by analyzing their mass spectra and using knowledge of chemistry."[2] Expert systems were limited by the need to enter and maintain rules manually; they were constrained to the vocabulary of the rules and did not scale to learn new domain knowledge automatically.

The second wave in AI hit by the end of the last century, and we are still riding it. It is signified by machine learning, where one trains a computational model on data, and upon deployment of the model in the real world, it can act on data it has never seen before. Moreover, the models learn from these interactions, thereby improving the model over time. The second wave is distinct from the first in that it is driven by the plethora of digital data, especially data that represents human behavior, for example, what our movie preferences are, when and how we commute to work every day, and what groceries we buy.

One form of machine learning models is deep neural networks (DNNs),[3] which are characterized by multiple layers of nodes, where each layer is connected to the next by weighted edges. Each node at each layer computes a function based on input weights and correspondingly outputs weights for nodes in the next layer. Overall, a DNN transforms input data (e.g., an image) into an abstract representation of the data (e.g., a classification of that image). As a simple example, suppose a deep neural network was trained to classify images. Then if we feed it a picture of a cat that it has never seen before, it will output that it is a cat. More precisely, it will output the classification label "cat" with an associated high probability, say 0.95, and perhaps a different label, for example, "tiger," with a lower probability, say 0.01. The DNN boom took off when, in 2012, AlexNet won the ImageNet contest on 1.2 million images and 1,000 classes.[4] DNNs showed how with Big Data and Big Compute, machines could perform certain tasks as well as or better than humans. They are part of computer vision systems in self-driving cars; they enable voice recognition in personal assistants on our phones and tabletop devices in our living rooms; and they were at the core of the computing system that in 2016 beat the best human Go player in the world.

Fast-forward ten years into the Age of Generative AI, where we can generate new data, such as text that has never been written before or images that have never been created before. One generative AI technique is based on large-language models (LLMs).[5] Authors of an early exploratory paper contrasting ChatGPT versus GPT4 show how GPT4 is able to generate a proof, which requires knowledge of calculus, to a simplification of a problem statement that appeared in the 2022 International Mathematics Olympiad.[6] Another generative AI technique is based on diffusion models,[7] which are especially good at generating images. A diffusion model first iteratively adds noise to the original picture, say of a dog, and then iteratively denoises to get a new image, that is, a brand-new picture of a dog.

To put science in the context of this AI time line, by the 1980s, supercomputers became the workhorse of science, for example, performing enormous numerical calculations and running complex simulations of physics-based models. The explosion of scientific data generated by devices and instruments enabled the age of data-driven science. From embedded microchips to space telescopes, scientists could sense the world, take measurements, record dynamics, and produce images of natural systems at unprecedented scale and speed.

At about the same time, statistical machine learning pulled the splintered subfields of AI back together, bringing vision, language, and robotics closer to each other, and even teasing us about the eventuality of artificial general intelligence (AGI).

It is the convergence of Big Compute, Big Data, and advanced AI that provides the context of this panel on using AI to make scientific discoveries. For science, the real breakthrough event came in 2018 with AlphaFold, an AI system built on both deep learning and reinforcement learning that could predict protein structure.[8]

In just the past 10 years, the second wave has turned into a tsunami. We have seen a 3.4-month doubling in the past 10 years in computational power (measured in petaflops per day) used to create machine learning models (Figure 3.1). In contrast, the 2-year doubling due to Moore's Law looks like the lower part of the curve. Note the y-axis is a log scale. Since 2010, we have also seen a 2.2× growth rate in training data size.[9] In just the past 4 years, we have gone from talking about billions of words with OpenAI's GPT3 to trillions with Databricks's DBRX.

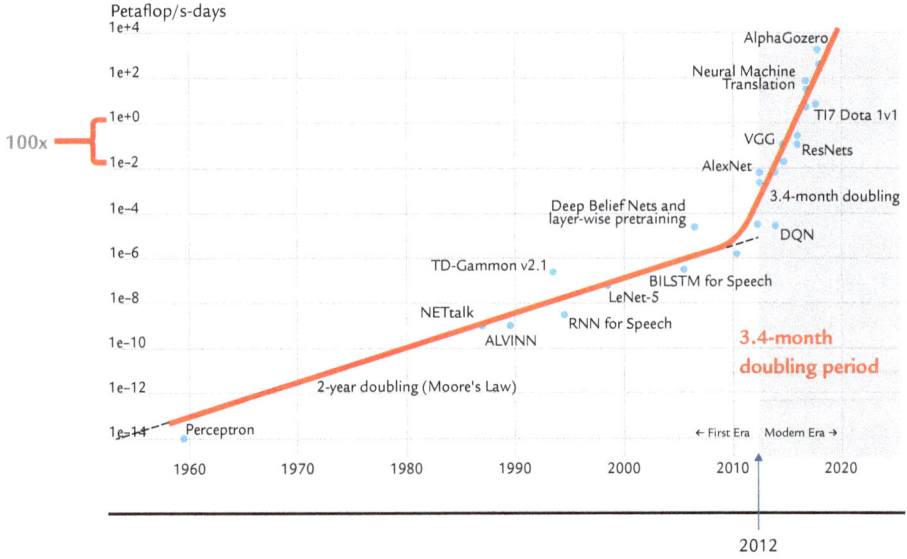

Figure 3.1. Compute versus machine learning model. Graphic from "AI and compute," OpenAI, May 16, 2018, https://openai.com/index/ai-and-compute/.

Part 2: Generative AI Architecture

How can scientists ride this tsunami? It is worth understanding a few basic concepts underlying generative AI, the cause for the recent disruption in the AI time line, and whose impact will be felt by all fields of endeavor for the long-term future. There is no going back. Today, generative AI is particularly remarkable for generating text and images. Tomorrow, who knows?

For example, to generate text,[10] if we feed a large-language model a sequence of words, then the LLM will predict the next word. That is all it does! It will produce the word with the highest probability of occurring next. More formally, when given an initial $i-1$ words, we draw the next word from the distribution of possible next words:

$$P(w_i \mid w_1, \ldots, w_{i-1})$$

For example, if we feed in the input sequence "The cute dog begged for a" it will output "bone" assuming no other word has a higher associated probability.

Example	Probability
The cute dog begged for a **bone**	0.85
The cute dog begged for a promotion	0.02

As mentioned earlier, we can generate new images using not just LLMs but diffusion models. The diffusion model was inspired by nonequilibrium statistical physics;[11] natural diffusion processes are found in physics, chemistry, and biology. In a diffusion model, the forward process systematically adds noise to an image; we then learn a reverse process by denoising, and in this process generate a new image (Figure 3.2).

What is fundamental to both techniques is that the underlying probability distributions are learned from billions of examples and represented as deep neural networks. Consider the transformer architecture,[12] which is shared by state-of-the-art large-language models. The transformer architecture has two steps, and both steps build on DNNs.[13] In the first step, through a series of transformations, we encode the input into a multidimensional embedding space (Figure 3.3).

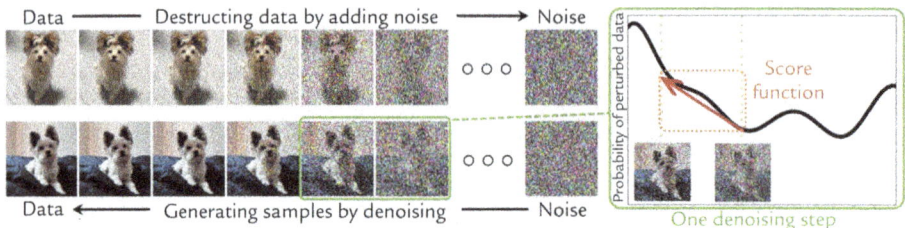

Figure 3.2. Generating a new dog image. Example from Lin Yang, Zhilong Zhang, Yang Song, Shenda Hong, et al., "Diffusion Models: A Comprehensive Survey of Methods and Applications," *ACM Computing Surveys* 56, no. 4 (November 9, 2023): 1–39, https://doi.org/10.1145/3626235.

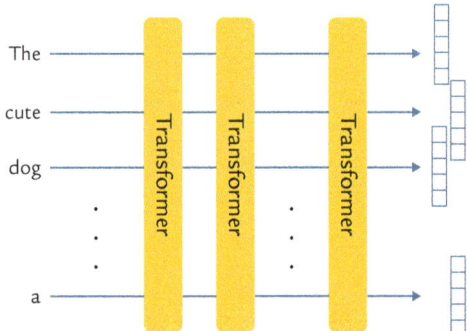

Figure 3.3. Transformer architecture, step 1: encode the input. Drawn from John Launchbury, "The Trajectory of AI," Presentation at Galois, Portland, OR, December 2023.

The manifold hypothesis states that real-world high-dimensional data lie on low-dimensional manifolds embedded in the high-dimensional space.[14] Thus, we can imagine that each transformer via linear and nonlinear operations stretches and squashes manifolds in this space, passing the transformed manifold onto the next layer. While we do not know if the manifold hypothesis is true, minimally it provides good intuition as to what happens at each layer. The result of this step is that each input word is embedded in this space. Embeddings make it possible to represent symbolic information numerically. Each word gets represented as a vector of floating point numbers, each of which represents some feature of the word (see

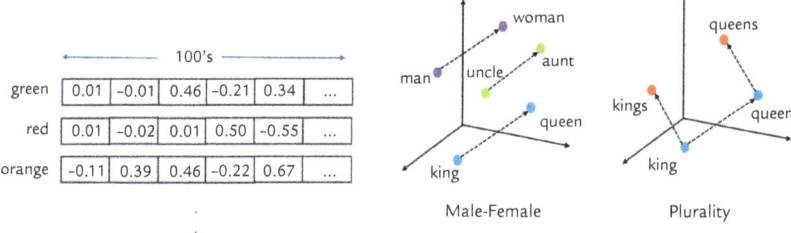

Figure 3.4. Embeddings represent knowledge abstractly.

left-hand side of Figure 3.4). Embeddings convert high-dimensional data into a low-dimensional space.

What is interesting about embedding spaces is that both distance and direction have meaning; hence we often use the term *vector embeddings*. From embeddings we can learn abstract concepts not explicitly represented. In the embedding shown in the right-hand side of Figure 3.4, "king" is to "queen" as "man" is to "woman" represents the abstract concept of gender. And "king" is to "queen" as "kings" is to "queens" represents plurality. (These examples are taken from Figure 2 in Mikolov, Yih, and Zweig 2013.[15])

In the second step, we decode the input sequence of words in the embedding space and eventually output the word with the highest probability of occurring next.[16] But unlike DNNs of the past, critical to the success of this architecture, we also add attention layers in between the transformer layers (Figure 3.5).

After all these transformations, we finally output "bone" (with an associated probability, say 0.85), which is then appended to the previous output tokens and used as input for the next iteration. This iterative process is how ChatGPT works.

Attention layers provide context for the words being processed. Context constrains the possibilities. Each node provides *key* information about itself to others, for example, "I'm a noun." And each node can *query* for information from its neighbors, for example, "I need a color." (*Key* and *query* are terms used in information retrieval, also used in Vaswani et al. 2017.[17])

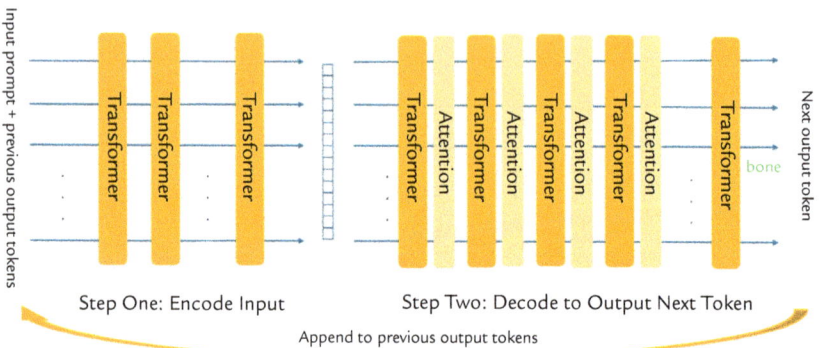

Figure 3.5. Transformer architecture, step two: decode to output next token; attention layers provide context. Drawn from John Launchbury, "The Trajectory of AI," presentation at Galois, Portland, OR, December 2023.

This context can help us resolve ambiguity in language. For example, Winograd Schema,[18] which are considered mini-Turing tests, are easily solvable by large-language models, and thus makes *it seem as if* LLMs can do common-sense reasoning. In this example,

> The trophy doesn't fit in the brown suitcase because it is too [large/small].

it is ambiguous whether "it" refers to "trophy" or "suitcase." We would know how to resolve the ambiguity if we know whether the last word is "large" or "small," which might be in the context of a longer input sequence.

Part 3: AI for Science

Let's take a step back from AI and explore how science can benefit from AI. Scientists can use generative AI to generate synthetic data, generate simulations, and more interestingly generate new hypotheses. The novelty of these new hypotheses is that because the computer has access to an enormous amount of data, it can find patterns or correlations that would never occur to a human or would take more than a lifetime for a human to uncover.

Even with more established techniques, such as DNNs, scientists can use AI for identification and discovery. We can classify and predict objects, recognize and discover new patterns, and detect anomalies and rare events. We can design and optimize experiments with AI recommending what control parameters to try for the next experiment. Automated experimental design is especially cost-effective and time-saving when running experiments on large, expensive instruments (e.g., a cyclotron, telescope, or neutrino detector). We could even use AI to propose new experiments and protocols to run. And it is a given that by using current AI techniques (e.g., LLMs), scientists can automatically pore through scientific literature, summarize results quickly, and create compelling visualizations. Finally, because AI techniques are agnostic as to what field of science we work in, by using AI we have the potential to expedite cross-disciplinary work.

Cutting across these categories of application, we can draw on a multitude of diverse sources of data to train and test new AI models for science:

- Scientific publications, preprints, lab notebooks
- Databanks, shared repositories, github
- Data from experiments
- Data from devices and scientific instruments (small to large)
- Data from simulations
- Data from the internet/web

These sources can be multimodal, structured and unstructured: text, images, graphs, tables, audio, video, clinical, software, and so forth.

To be more specific, consider two scientific disciplines, not covered by the panelists, to see how AI has already been helping to make new discoveries.

In astronomy,[19] scientists used DNNs to recognize galaxies and now can classify galaxies with an accuracy of 98 percent. Astronomers used AI to detect new exoplanets, to predict signatures of new types of gravitational waves, and to find a unique object that may be a remnant of two black holes merging. They used generative AI to produce a sharper image of the very first image of a black hole.

In materials science,[20] the design space is huge. A short polymer with 100 amino acids has on the order of 10^{130} designs, more than the number of atoms in the universe. Materials scientists are using generative AI to create new material designs. For example, Markus Buehler used LLMs to create a never-before-seen design of a hierarchical mycelium-based composite.[21] Materials scientists are exploring how to use AI to identify new equations and algorithms, to synthesize complex novel proteins that do not exist in nature, to visualize complex systems, and to predict how a new material will behave.

What's in the Future?

Looking ahead, already the scientific community is exploring how to build foundational models for their scientific domain, which can be later fine-tuned to a specific problem or even to other domains. For example, Shirley Ho of the Flatiron Institute is leading the Polymathic AI initiative, an international and multidisciplinary team of collaborators, including experts from physics, astrophysics, mathematics, artificial intelligence, and neuroscience, to build foundational models that could be applied to a wide range of scientific problems.[22] As another example, Prov-GigaPath is a whole-slide pathology foundation model pretrained on 1.3 billion image tiles in 171,189 whole slides from Providence, a large US health network.[23]

Scientists can tailor concepts from the transformer architecture model and apply them to their domain. What is the analogy to predicting the next word? What are analogies to abstractions such as language, grammar, embeddings, and context? For example, a collaboration between astronomers and computer scientists are exploring "planetary linguistics" to determine whether planetary systems fall into natural categories following grammatical rules.[24]

Although Big Data and Big Compute have been responsible for driving the Second Wave of AI, for many reasons, the scientific community, including computer science and AI researchers, should pursue what can be done with Small Data and Small Compute.[25] Currently only those working in a handful of big technology companies have access to the large amounts of data and compute to train and build state-of-the-art AI models; the academic community is impoverished. Can we get similar functionality with less, perhaps through cleverer algorithms? Moreover, we may not have an abundance of data in some scientific domains. Finally, building today's models incurs enormous energy usage; building smaller models with less data could be more energy-efficient.

One direction the AI community could pursue is combining symbolic models of the past with statistical models of today.[26] A different hybrid approach for science is to combine machine learning with physics-based models (e.g., for simulations). For example, one aim of the National Science Foundation (NSF) Science and Technology Center "Learning the Earth with AI and Physics (LEAP)" is to reduce the uncertainty envelopes of climate model predictions using machine learning.[27]

There are some challenges, the first of which is having enough reliable scientific data.[28] And finally, AI raises a new challenge to ensuring scientific integrity. To address this challenge, we not only need to educate scientists to check the accuracy of AI outputs[29] but also to do more research on trustworthy AI.[30]

AI has been around for decades, but today's AI craze has captured the fascination of the public and media. Is this another technology fad? Definitively not. The next generation will not know a world without generative AI as part of their lives, much like the current generation without the internet or smartphones. How should the scientific community respond? Understand it, embrace it, and explore with it.

Acknowledgments

My contribution on this topic is primarily in the compilation of ideas from the literature and from content in talks of others (see footnotes). I am grateful to Chris Impey for his permission to use content from his paper on AI and astrophysics, and to John Launchbury and Rebecca Willette for their permission to use content from their presentations for my talk and for this article. Additionally, I borrowed heavily from the talk by Markus Buehler in my discussion on AI and materials science.

Notes

1. Robert K. Lindsay, Bruce G. Buchanan, Edward A. Feigenbaum, and Joshua Lederberg, "DENDRAL: A Case Study of the First Expert System for Scientific Hypothesis Formation," *Artificial Intelligence* 61, no. 2 (1993): 209–261, https://doi.org/10.1016/0004-3702(93)90068-M.

2. "Dendral," *Wikipedia* (n.d.), https://en.wikipedia.org/wiki/Dendral.

3. Yann LeCun, Yoshua Bengio, and Geoffrey Hinton, "Deep Learning," *Nature* 521 (May 27, 2015): 436–444, https://www.nature.com/articles/nature14539.

4. Alex Krizhevsky, Ilya Sutskever, and Geoffrey E. Hinton, "ImageNet Classification with Deep Convolutional Neural Networks," *Communications of the ACM* 60, no. 6 (May 24, 2017): 94–90, https://doi.org/10.1145/3065386.

5. Ashish Vaswani, Noam Shazeer, Niki Parmar, Jakob Uszkoreit, et al., "Attention Is All You Need," *Proceedings of the 31st International Conference on Neural Information Processing Systems* (December 4, 2017): 6000–6010.

6. Sébastien Bubeck, Varun Chandrasekaran, Ronen Eldan, Johannes Gehrke, et al., "Sparks of Artificial General Intelligence: Early Experiments with GPT-4," *arXiv*, April 2023, arXiv:2303.12712.

7. Jascha Sohl-Dickstein, Eric Weiss, Niru Maheswaranathan, and Surya Ganguli, "Deep Unsupervised Learning using Nonequilibrium Thermodynamics," *Proceedings of the 32nd International Conference on Machine Learning* (2015), https://proceedings.mlr.press/v37/sohl-dickstein15.html.

8. John Jumper, Richard Evans, Alexander Pritzel, Tim Green, et al., "Highly Accurate Protein Structure Prediction with AlphaFold," *Nature* 596 (July 15, 2021): 583–589, https://www.nature.com/articles/s41586-021-03819-2.

9. Cade Metz, Cecilia Kang, Sheera Frenkel, Stuart A. Thompson, et al., "How Tech Giants Cut Corners to Harvest Data for AI," *New York Times*, April 6, 2024, https://www.nytimes.com/2024/04/06/technology/tech-giants-harvest-data-artificial-intelligence.html.

10. The dog LLM example is from Rebecca Willett's talk at CCC@AAAS 2024 (Part One): "Generative AI in Science: Promises and Pitfalls."

11. Sohl-Dickstein et al., "Deep Unsupervised Learning using Nonequilibrium Thermodynamics."

12. Vaswani et al., "Attention Is All You Need."

13. The high-level depiction of the transformer architecture and the embedding example are drawn from John Launchbury's "The Trajectory of AI" talk on December 1, 2023: https://galois.com/blog/2023/12/the-trajectory-of-ai/.

14. "Manifold Wikipedia," *Wikipedia* (n.d.), https://en.wikipedia.org/wiki/Manifold_hypothesis.

15. Tomas Mikolov, Wen-tau Yih, and Geoffrey Zweig, "Linguistic Regularities in Continuous Space Word Representations," *Proceedings of NAACL-HLT* (2013): 746–751, https://aclanthology.org/N13-1090.pdf.

16. LLMs technically operate over "tokens" not "words"; one can think of a "token" as a sequence of contiguous characters, including letters, punctuation, and perhaps other delimiters. In this paper, we use "word" and "token" interchangeably.

17. Vaswani et al., "Attention Is All You Need."

18. Terry Winograd, "Understanding Natural Language," *Cognitive Psychology* 3, no. 1 (January 1972): 1–191, https://doi.org/10.1016/0010-0285(72)90002-3.

19. All astronomy examples are from Chris Impey, "AI Is Helping Astronomers Make New Discoveries and Learn About the Universe Faster than Ever Before," *The Conversation*, May 3, 2023, https://theconversation.com/ai-is-helping-astronomers-make-new-discoveries -and-learn-about-the-universe-faster-than-ever-before-204351.

20. The polymer example and exploratory ideas for materials science are from a talk by Markus Buehler at CCC@AAAS 2024 (Part Two): "Generative AI in Science: Promises and Pitfalls."

21. Markus J. Buehler, "Accelerating Scientific Discovery with Generative Knowledge Extraction, Graph-Based Representation, and Multimodal Intelligent Graph Reasoning," *arXiv*, June 10, 2024, https://arxiv.org/abs/2403.11996.

22. Michael McCabe, Bruno Régaldo-Saint Blancard, Liam Holden Parker, Ruben Ohana, et al., "Multiple Physics Pretraining for Physical Surrogate Models," *arXiv*, October 4, 2023, https://arxiv.org/abs/2310.02994.

23. Hanwen Xu, Naoto Usuyama, Jaspreet Bagga, Sheng Zhang, et al., "A whole-slide foundation model for digital pathology from real-world data," *Nature* 630 (May 22, 2024): 181–188, https://doi.org/10.1038/s41586-024-07441-w.

24. Emily Sandford, David Kipping, and Michael Collins, "On Planetary Systems as Ordered Sequence," *Monthly Notices of the Royal Astronomical Society* 505, no. 2 (August 2021): 2224–2246, https://doi.org/10.1093/mnras/stab1480.

25. Jeannette M. Wing and Michael Wooldridge, *Findings and Recommendations of the May 2022 UK-US AI Workshop* (National Science Foundation and Engineering and Physical Sciences Research Council, May 3–4, 2022), https://www.cs.columbia.edu/~wing /publications/WingWooldridge2022.pdf.

26. Wing and Wooldridge, "Findings and Recommendations of the May 2022 UK-US AI Workshop."

27. "About Learning the Earth using Artificial Intelligence and Physics (LEAP)," National Science Foundation Science and Technology Center Program, 2021, https://leap .columbia.edu/about/.

28. Jennifer Listgarten, "The Perpetual Motion Machine of AI-Generated Data and the Distraction of ChatGPT as a 'Scientist,'" *Nature Biotechnology* 42 (January 25, 2024): 371–373, https://doi.org/10.1038/s41587-023-02103-0.

29. Wolfgang Blau, Vinton G. Cerf, Juan Enriquez, Joseph S. Francisco, et al., "Protecting Integrity in the Age of Generative AI," *Proceedings of the National Academy of Sciences* 121, no. 22 (May 2024).

30. Jeannette M. Wing, "Trustworthy AI," *Communications of the ACM* 64, no. 10 (October 2021): 64–71; David "davidad" Dalrymple, Joar Skalse, Yoshua Bengio, Stuart Russell, et al., "Towards Guaranteed Safe AI: A Framework for Ensuring Robust and Reliable AI Systems," *arXiv*, May 10, 2024, https://arxiv.org/abs/2405.06624.

CHAPTER 4

We've Been Here Before: Historical Precedents for Managing Artificial Intelligence

Marc Aidinoff and David I. Kaiser

Introduction

Scientific and technological innovations are made by people, and so they can be governed by people. Notwithstanding breathless popular descriptions of disempowered citizens cowed by technical complexity or bowing to the inevitable march of the new, history teaches that novel technologies like artificial intelligence can—indeed, must—be developed with ongoing and meaningful democratic oversight. Self-policing by technical experts is never enough to sustain an innovation ecosystem worthy of public trust. Contemporary artificial intelligence (AI) and related computing techniques might be distinct technological phenomena, but they too can be governed in the public interest.

Rather than treat AI governance as an abstract problem, we urge policymakers to rely on the rich, empirical record of past engagements to conceptualize and respond appropriately to present-day challenges. History offers a repository of multiple, overlapping, real-world instances in which technical experts, policymakers, and broader publics have grappled with once-new technologies. Commentators and policymakers too often focus narrowly on one historical episode or analogy when thinking about the challenges of novel technologies—most commonly turning to the sprawling

Manhattan Project during the Second World War or life scientists' famous meeting at Asilomar in the mid-1970s. Yet considering *multiple* analogies and disanalogies can elucidate complementary axes along which to assess likely harms and potential benefits.

In this brief paper, we consider three historical episodes: the early nuclear weapons complex during the 1940s and 1950s; biotechnology, biomedicine, and the implementation of various safeguards in the 1970s; and the adoption and oversight of forensic technologies within the US legal and criminal justice systems over the course of the past century. Each example offers distinct insights for understanding opportunities and risks associated with AI today. As we discuss, each of the past examples required a broad range of actors to think at different scales: national and global security, the health of local communities, and individuals' civil rights. No example offers a perfect analogy with present-day challenges; yet even the disanalogies can help clarify realistic options for decision-making today.

As each of the previous historical episodes make clear, the scientific and technical communities have often taken on special roles in establishing norms regarding how to define and protect the public interest. Yet in none of these previous instances did scientists and technologists hold unilateral sway over how the new technologies would be assessed, deployed, or governed. History offers the opportunity to consider how each previous effort succeeded in some ways but fell short in others. Across each example, we therefore identify three key themes for thinking about the governance of AI today: *the inadequacy of researchers' self-policing* to produce meaningful safeguards on impactful technologies that move beyond controlled laboratory settings; *the necessity of broad-gauge input and oversight* to sustain an innovation ecosystem; and finally, *the need for recurring reviews* to regularly reassess evolving technologies and the shifting social practices within which they are embedded.

Part 1: Nuclear Secrets

The path from basic discoveries in nuclear science to sprawling weapons programs was dizzyingly short. The first indication of nuclear fission caught chemists Otto Hahn and Fritz Strassmann by surprise in their Berlin laboratory in December 1938. Immediately upon receiving an update from Hahn by letter, the recently exiled theoretical physicist Lise Meitner and her nephew, Otto Robert Frisch, developed a remarkable interpretation: Under certain circumstances, bombardment of a heavy nucleus such as uranium by neutrons could split the nucleus and release additional energy.[1] Hahn, Strassmann, Meitner, and Frisch each communicated their results in rapid-fire scientific publications as well as via informal discussions with colleagues; within weeks, scientists around the world began pursuing follow-up studies. Frisch's mentor, Niels Bohr, teamed up with another protégé, American physicist John Wheeler, to produce a detailed theoretical analysis of nuclear fission. Their landmark article was published in the *Physical Review* on September 1, 1939, just as Nazi tanks invaded Poland, triggering the start of the Second World War.[2]

Even before the Bohr–Wheeler paper had been published, scientists in at least five countries had recognized the possibility that nuclear fission could be used to create a new type of weapon and had initiated discussions with government officials. In April 1939, the German Reich Ministry of Education held a secret meeting on military applications of nuclear fission and banned uranium exports. That same month, the Japanese government launched "Project Ni" to study possible weapons effects of fission, while, independently, several physicists in Britain urged their government to jumpstart a nuclear weapons project by securing uranium ore from the Belgian Congo. In August 1939, Albert Einstein signed a letter to US President Franklin Roosevelt—which had been written by concerned émigré physicists Leo Szilard and Eugene Wigner—alerting Roosevelt of the possibility that nuclear weapons could exploit runaway fission chain reactions. A few weeks later, Leningrad physicist Igor Kurchatov informed the Soviet government about possible military applications of nuclear fission.[3]

Given the plausible connections between nuclear fission and new types of weapons—and set against the drumbeat of worsening international relations—some scientists sought to control the flow of information about nuclear fission. Beginning in spring 1939, Hungarian physicist Leo Szilard, who had fled Europe and landed in New York City, urged his colleagues to adopt a voluntary moratorium on publishing new results. When some physicists refused to withhold their latest findings, Szilard concocted a new plan to allow researchers to submit their articles to scientific journals—which would enable clear cataloging of priority claims—but coordinate with the journal editors to hold back publication of certain papers until their release could be deemed safe. This scheme, too, proved difficult to implement in practice, not least because it depended upon voluntary compliance, with no means of enforcement.[4] It also had some unintended consequences. When Kurchatov and his colleagues in the Soviet Union noticed a distinct falloff of publications in the *Physical Review* regarding nuclear fission, they considered their suspicions confirmed and doubled down on their efforts to convince Soviet officials to take the matter seriously.[5]

Szilard's proposals focused on controlling the flow of information rather than regulating research itself. That distinction disappeared once the Allied efforts on nuclear weapons became more formalized, scaling up from lackluster study groups to the Manhattan Engineer District in June 1942. Under the auspices of the newly formed Office of Scientific Research and Development (OSRD) and administered by the US Army Corps of Engineers, officials in the Manhattan Project imported older procedures for military secrecy and provisioning—some dating to the 1917 US Espionage Act, enacted in a hurry after the United States had entered the First World War—to exert control over the circulation of information, materials, and personnel. The US Federal Bureau of Investigation (FBI) and the Military Intelligence Division conducted background checks on researchers; General Leslie Groves imposed strict compartmentalization rules to try to limit how much information any single individual could glean about the sprawling project; massive infrastructure was devoted to producing fissionable materials within secret facilities at places like Oak Ridge, Tennessee, and Hanford, Washington; while more mundane materials, such as rubber and gasoline—by then under strict wartime rationing—were diverted to the high-priority project.[6]

Over the course of the war, older conventions regarding secrecy and classification were updated and specialized to the case of nuclear weapons. These newer routines were formalized with passage of the US Atomic Energy Act in August 1946. Although the Act transferred control over the nuclear complex from the War Department to a new civilian agency—the Atomic Energy Commission (AEC)—in many ways the AEC reinforced wartime procedures. Under the new law, for example, whole categories of information about nuclear science and technology were deemed to be "born secret," that is, classified by default and only released following careful review. The Act also established a government monopoly over the development and circulation of various fissile materials within the US, effectively foreclosing efforts by private companies to pursue civilian nuclear power generation. (Several of these provisions of the Act were amended in 1954, with the explicit goal of fostering private-sector efforts in nuclear power, but with mixed results.[7])

Policymakers crafted these regulatory developments amid specific domestic and international considerations. On the international front, mutual suspicions between officials in the United States and the Soviet Union—exacerbated by the shocking revelation in February 1946, following the defection of a Soviet cipher clerk, that the Soviets had conducted espionage at several Manhattan Project sites during the war—derailed early efforts to establish international control of nuclear science and technology. Domestically, long-standing rivalries between various military branches shaped debates over nuclear weapons policies, including whether the United States should pursue next-generation weapons such as thermonuclear (or fusion) bombs.[8]

Much as Szilard had done as early as 1939, after the war many scientists and engineers worked hard to help shape the evolving landscape of practices and norms around nuclear science and technology. Some, like J. Robert Oppenheimer, moved from leadership positions in the wartime program into influential consulting roles after the war. Oppenheimer helped draft several proposals for postwar nuclear policies and chaired the new General Advisory Committee of the AEC. Others, especially younger colleagues, formed new organizations like the Federation of Atomic Scientists to lobby lawmakers for their preferred policy outcomes, such as civilian (rather than military) control of the postwar nuclear complex, and in support of nuclear disarmament.[9]

Before long, however, the scientists' illusions of control collapsed amid Cold War realities. Right on the heels of their major legislative victory—ensuring passage of the Atomic Energy Act that enshrined civilian oversight—groups like the Federation of Atomic Scientists became targets of a concerted campaign. The FBI and the US House Committee on Un-American Activities targeted the Federation and several of its individual members, smearing them with selective leaks and high-profile hearings, alleging Communist sympathies.[10] Oppenheimer's infamous hearing in June 1954 before an AEC personnel security board was a late example of what had long since become routine. In fact, a disproportionate number of younger, more vulnerable nuclear physicists were affected by domestic anti-communism than representatives of any other academic discipline during the decade after the end of the Second World War. The elaborate system of nuclear classification became a cudgel with which to silence critics, whose attorneys were often denied access to information under the guise of protecting national security.[11]

Beyond the impact on individuals and groups, the postwar nuclear classification regime strained relationships with US allies—most notably the United Kingdom—while remaining relatively ineffective at halting nuclear proliferation. Within a few years after the war, the Soviet Union built both fission and fusion bombs with a speed that caught many US authorities off guard; those efforts were aided, in part, by wartime espionage that had pierced military control. Arguably, overzealous efforts at nuclear secrecy helped to accelerate the arms race, exacerbating the precarious brinksmanship of a protracted Cold War and triggering all-too-hot proxy wars around the globe.[12]

As policymakers ask questions today about allowing researchers to deploy, withhold, or partially disclose new computational models and techniques, the example of nuclear secrecy infrastructure provides important cautions about bureaucratic overreach and political abuse. During the postwar years, few scientists, engineers, or policymakers suggested that *all* information about nuclear weapons or related technologies should be openly shared—proliferation concerns were real and some safeguards were clearly appropriate. Yet the complex system of nuclear classification and control quickly grew so byzantine that legitimate research inquiries were cut off, responsible private-sector investment was stymied, and open political debate was squashed.[13] As the secrecy regimes grew in complexity and extensiveness, the academic community often served as a weak but crucial counterbalance to maintain, or at least seek to maintain, the levels of openness necessary for robust scientific progress and democratic oversight.

Part 2: Biotechnology and Biomedicine

Leo Szilard's first impulse, upon learning about nuclear fission in 1939, had been to try to convince his fellow scientists to adopt a voluntary moratorium on publishing certain findings. Several decades later, in the mid-1970s, a group of molecular biologists followed a similar route, urging their colleagues to pause research involving the new techniques of recombinant DNA (rDNA). The call by Stanford biologist Paul Berg, together with colleagues from several other elite US universities and research sites, moved beyond Szilard's earlier intervention: They pressed for a voluntary moratorium on certain types of research, not only on publication.[14]

By the spring of 1974, Berg and his colleagues had grown concerned about potential risks of rDNA research, even as they anticipated many beneficial outcomes. What if pathogenic bacteria acquired antibiotic-resistant genes, or carcinogenic genes were transferred to otherwise harmless microorganisms? Unlike the massive, top secret industrial sites of the wartime Manhattan Project, rDNA experimentation involved relatively small-scale, benchtop apparatus, and hence could be pursued within nondescript laboratories in urban centers—such as at Berg's and colleagues' universities. What types of containment facilities and safety protocols could protect researchers as well as their neighbors from possible leaks of dangerous biological materials? How could the risks of various research projects be assessed and mitigated?[15] As MIT's David Baltimore recalled soon after Berg and colleagues met in his office to brainstorm about their concerns, "we sat around for the day and said, 'How bad does the situation look?' And the answer that most of us came up with was that . . . just the simple scenarios that you could write down on paper were frightening enough that, for certain kinds of limited experiments using this technology, we didn't want to see them done at all."[16] Berg, Baltimore, and their small group published a brief, open letter calling for a voluntary moratorium on rDNA research—it appeared in *Science, Nature,* and the *Proceedings of the National Academy of Sciences*—until the scientific community could address such concerns.[17]

By the time their letter appeared in print, the Berg group had been deputized by the US National Academy of Sciences to convene a meeting of colleagues and develop recommendations for the US National Institutes of Health (NIH). Famously, that meeting was held in February 1975 at the Asilomar Conference Grounds in Pacific Grove, California. Berg, Baltimore, and their original discussion mates were joined by other eminent biologists, including Maxine Singer and Sydney Brenner. Much like the group that had met at MIT the previous spring, the Asilomar group consisted almost entirely of researchers in the life sciences.[18] They recommended a temporary extension of the voluntary research moratorium combined with a framework for assessing risks and appropriate containment facilities for various types of rDNA experiments. In late June 1976, the US Department of Health, Education, and Welfare released the official guidelines that would govern rDNA research by NIH-funded researchers throughout the United States, which drew extensively upon the Asilomar recommendations.[19]

To this day, the Asilomar meeting is routinely hailed as the preeminent example of how scientists can successfully and responsibly govern risky research: Concerned scientists spoke up, urged restraint upon their colleagues, and forged new guidelines among themselves. Yet much like Szilard's calls for nuclear scientists to self-censor during the early days of nuclear fission, the biologists' self-policing around rDNA was a small part of what grew into a much larger process—one that involved input and negotiation among a much wider set of stakeholders.[20] On the very evening in June 1976 that federal officials announced the new NIH guidelines, the mayor of Cambridge, Massachusetts—home to famously difficult-to-govern research institutions like Harvard University and MIT—convened a special Hearing on Recombinant DNA Experimentation. As Mayor Alfred Vellucci announced upon opening the special session, "No one person or group has a monopoly on the interests at stake. Whether this research takes place here or elsewhere, whether it produces good or evil, all of us stand to be affected by the outcome. As such, the debate must take place in the public forum with you, the public, taking a major role."[21] And so began a remarkable months-long effort by local university researchers, private-practice physicians, city officials, and other concerned citizens to devise an appropriate regulatory framework that would govern rDNA research within Cambridge city limits—under threat of a complete ban if the new Cambridge Experimentation Review Board (CERB) failed to converge on rules that could pass muster with the city council.[22]

The CERB group held open, public meetings twice weekly throughout the autumn of 1976. During the sessions, Harvard and MIT researchers had opportunities to explain details of their proposed research to nonspecialists; on other evenings, CERB hosted public debates over proposals for competing safety protocols. Similar civic groups met to hash out local regulations in cities across the United States, including Ann Arbor, Michigan; Bloomington, Indiana; Madison, Wisconsin; Princeton, New Jersey; as well as Berkeley and San Diego in California. In none of these jurisdictions did citizens simply adopt the scientists' Asilomar recommendations without thorough discussion, scrutiny, and debate. For example, the CERB group called for the formation of a new five-person Cambridge Biohazards Committee plus regular site inspections of rDNA labs within city limits, exceeding the requirements of the federal NIH guidelines. Only after CERB's extensive, at times thorny, negotiations did the Cambridge city council vote unanimously, in early February 1977, to adopt the locally written Ordinance for the Use of Recombinant DNA Molecule Technology within the city—two years after the Asilomar meeting.[23]

With the carefully negotiated Cambridge ordinance in place, the city quickly became a biotechnology juggernaut, earning the nickname "Genetown." City officials, university administrators, laboratory scientists, and neighboring nonscientists had worked together to construct a clear regulatory scheme within which new types of scientific research could thrive—both within university settings and quickly within spin-off biotech companies as well.[24] The extended effort of public participation and debate helped to establish a new level of public trust, while avoiding Manhattan Project–style monopolies.

In parallel with the rDNA efforts, biomedical researchers, policymakers, and regulators across the United States forged a separate regulatory framework during the 1970s, which likewise required life scientists to work closely with concerned nonscientists. Following headline-grabbing revelations of egregious abuses of participants in previous biomedical studies—including the long-running Tuskegee Syphilis Study on Black men in rural Alabama—the US Congress passed the National Research Act in 1974. The Act stipulated the creation of a new national commission that would recommend uniform requirements to protect individuals who were involved in research studies.[25]

In 1979, the commission published the *Belmont Report*, articulating general principles and specific practices regarding the treatment of "human subjects" in federally funded research. Among the new requirements: ensuring that participants in research studies granted "informed consent," and that potential risks to individual participants were appropriately balanced by potential benefits of a given study. To evaluate and oversee such requirements, the National Research Act codified that federally funded research involving human subjects must be reviewed by a local "institutional review board," or IRB, whose membership had to include individuals with a range of experiences and expertise. At least one member of each IRB had to represent "nonscientific" concerns.[26]

Much like the CERB process in Cambridge, the National Research Act required biomedical researchers—at least those working with federal funds—to negotiate safe and effective research practices, with input and oversight extending beyond the research community itself. Imperfect and at times frustratingly bureaucratic, the new IRB infrastructure did not force all research to grind to a halt. Rather, it formalized a set of practices that had been honed within NIH's own research centers to mitigate real harms.[27]

The 1974 National Research Act and the 1979 *Belmont Report* were forged in response to specific concerns at the time. Although the so-called Common Rule (US federal law 45 C.F.R. 46) which governs research on human subjects has been updated as recently as 2017, the current provisions still do not map effectively to more recent forms of research involving human-sourced data and information, especially those for which potential harms need not arise at the point of data collection.

As technical systems come to depend on more and more sensitive data, these regulatory regimes are clearly insufficient, especially when researchers are separated from data collection by relying on third-party vendors.[28] Few if any individuals have granted consent (informed or otherwise) for their personal data, medical records, or facial images to be used as training data for such massive algorithmic projects. Likewise, although the Common Rule includes clear definitions of "identifiable private information" that is to be protected for study subjects, recent computational projects that rely upon amassing and analyzing large datasets routinely violate stated privacy protections, even when manipulating "deidentified" datasets.[29]

Part 3: Forensic Science

Whereas scientists like Leo Szilard and Paul Berg tried to quickly craft guardrails around the scientific work they were developing, John Larson was eager to deploy his latest innovation: the cardio-pneumo-psychograph device, or "polygraph." Larson's goal was not new; uncovering submerged human truths had long been a goal of physiological inquiry. Nineteenth-century physicians were particularly interested in the way the body could betray the mind. Étienne-Jules Marey, for example, took physical measurements of small changes to reveal stress, with the conviction that such measurements could reveal a hidden interior truth.[30] By the early twentieth century, leading psychologists were working to operationalize the emerging consensus that emotions were bodily. In 1917, William Moulton Marston and his wife Elizabeth Holloway Marston invented a form of the polygraph, but within a few years, Larson had added two crucial insights. The first was to take continuous measurements of blood pressure and record them as one running line, monitoring change relative to a baseline. The second was to partner with law enforcement.[31]

In the spring of 1921, Larson tried out his technology to solve a real crime, a potboiler-style drama of a missing diamond presumed stolen by one of ninety women living in a boarding house. The thief, whose recorded blood pressure did drop precipitously during her interrogation, eventually confessed after days of additional interrogation. With journalists eager for gripping narratives about the latest crime, the cardio-pneumo-psychograph made great copy, but to Larson's chagrin, it was renamed the "lie detector."

Historical accounts even credit newspapers with pressuring police in other jurisdictions to further adopt the tool. For August Vollmer, the chief of police in Berkeley, California, the cardio-pneumo-psychograph was particularly appealing because it could help professionalize law enforcement. Concerned with perceptions of a corrupt police force that relied on personal relations and intuitions, Vollmer was enthusiastic to experiment with new "scientific" policing. Although the methods were unproven, Vollmer believed that the patina of scientific expertise gained by enrolling Larson would bolster public support for local law enforcement.[32]

From the beginning, the polygraph was a "charismatic" technology that captured public interest.[33] It inspired popular depictions that led to the polygraph's widespread deployment beyond routine police work or formal legal settings. Some of these uses were relatively banal, such as trying to understand what drew certain audiences to films or actors. But the stakes of this unreliable technology grew in more impactful domains like employment. For example, adherence to the Cold War nuclear secrecy regime was policed through polygraph tests for adjudicating and maintaining security clearances. Beyond the nuclear complex, employers saw the polygraph as a useful screen for job suitability, despite its unreliability and recurring biases.[34]

Judges were less willing to accept the polygraph starting in 1922, with the trial of James Frye. Frye had previously confessed to the murder in question but claimed that his confession had been coerced. William Marston performed a polygraph test to validate Frye's claim. After a cursory review, the judge rejected the polygraph as evidence. The subsequent "Frye Rule" was designed to prevent scientific developments that were still undergoing development from entering the courtroom. Instead, a methodology like using polygraph machines would require "general acceptability" by the scientific community. This standard encoded a belief that juries would be distinctly swayed by supposedly objective scientific evidence produced by a machine like the polygraph.[35]

In practice, the Frye Rule did not prevent deeply questionable evidence from entering court proceedings—let alone from circulating beyond formal legal settings. In 1993, *Daubert v. Merrell Dow Pharmaceuticals* offered a new standard to replace the Frye Rule by further empowering judges to act as gatekeepers of expert testimony about novel technologies. It asked judges to think like scientists who evaluated peer-reviewed expertise. The goal remained the same: The courts would not be a place for radical experimentation with novel technologies.[36] (As recently as December 2023, amendments to the US Federal Rule of Evidence 702 have aimed to clarify that expert testimony in trial is to be treated as the expert's opinion, and that the proponent of introducing such testimony must meet a burden-of-evidence standard for such testimony to be admissible.[37])

Either relying on their own judgment or assessing the consensus views, judges needed to assess basic validity claims about lie detection. In turn, the scientific community repeatedly mobilized to limit the use of polygraphs in court. The US Office of Technology Assessment (OTA) concluded in a 1983 report that there was "only limited scientific evidence for establishing the validity of polygraph testing."[38] Again, the discrepancy between criminal law within a courtroom and deployment of the technology in other high-stakes arenas remained stark. The resistance to the polygraph from multiple experts in the most regulated legal sphere of criminal law was matched by an unchecked spread of the technology in other important spheres. The same OTA report estimated that outside of the federal government, more than one million polygraph tests were administered annually within the United States just to determine employment.[39] More recently, the US National Academies led efforts to (again) scrutinize evidence on the reliability of the polygraph.[40] The 2003 report has played a crucial role in keeping the polygraph out of courtrooms.

In contrast to polygraph evidence, other science-based techniques have long been incorporated within legal proceedings in the United States, such as fingerprint analysis. Although far from perfect, the use of fingerprint identification techniques within law enforcement and legal settings has been subject to expert review, training, and standardization for decades.[41] Moreover, high-profile misidentifications—such as the one in 2004 that led to the wrongful imprisonment of an American lawyer living in Oregon on charges related to the terrorist bombing of commuter trains in Madrid, Spain—catalyzed multiple reviews by expert panels to reassess the underlying scientific bases for fingerprint identifications and to update best-practice procedures for their use, including new types of training for practitioners.[42]

Algorithmic facial recognition technology has followed a trajectory more like the polygraph than like fingerprinting. Despite its significant, well-documented flaws, facial recognition technology has become ubiquitous in high-stakes contexts outside the courtroom.[43] A few years ago, the US National Institute of Standards and Technology (NIST) conducted a detailed evaluation of nearly 200 distinct facial recognition algorithms, from around 100 commercial vendors. Nearly all of the machine-learning algorithms demonstrated enormous disparities, yielding false-positive rates more than 100 times higher when applied to images of Black men from West Africa compared to images of white men from Eastern Europe; the NIST tests also found systematically elevated false-positive rates when applied to images of women than men across all geographical regions.[44] In the face of such clear-cut biases, some scholars have called for increased inclusion in the datasets—in theory broadening the types of faces that can be recognized.[45] Others have argued that inclusion is the problem rather than the solution, and that the imperative to include more data puts identified and misidentified citizens at increased risk, without legal recourse.[46]

Although the research community has identified these stark demographic biases and thousands of research papers have focused on ways to mitigate such disparities under pristine, laboratory conditions, the commercially available algorithms have already moved well beyond research spaces and into impactful real-world settings.[47] Within the United States alone, thousands of distinct law enforcement jurisdictions can purchase commercial facial recognition technologies, subject to no regulation, standardization, or oversight. This free-for-all has led to multiple reports of Black men being wrongfully arrested due to a combination of failures, for example, inadequate technical calibrations for the various algorithms together with human failures to follow recommended procedures following a putative facial image match within an active police investigation, such as seeking additional eyewitness testimony or forensic evidence from the crime scene.[48]

These continuing real-world failures—which exacerbate long-standing inequities within existing institutional frameworks—are likely to worsen in the absence of any oversight or regulation.[49] There already exist more than one billion surveillance cameras across fifty countries. Within the United States alone, facial images of half the adult population are already included in databases accessible to law enforcement.[50]

Like the polygraph, these faulty unregulated technologies have already moved far beyond both laboratory and law enforcement settings. In some cases, they have generated sensational claims that far outstrip technical feasibility, such as a widely covered 2018 study that claimed that algorithmic analysis of facial images could determine an individual's sexual orientation.[51] Meanwhile, private vendors continue to scoop up as many facial images as they can, almost always from platforms for which the people depicted neither granted permission for such uses nor were aware of the third-party data collection.[52] In turn, facial surveillance is now used to surveil all sorts of new contexts, including monitoring students' behavior in school, preventing access to venues, and even screening for jobs.[53]

Conclusions

AI policy is marked by a recurring problem: a sense that AI itself is difficult or even impossible to fully understand. Scholars have shown how machine learning relies on several forms of opacity: corporate secrecy, technical complexity, and unexplainable processes.[54] Scientists have a special obligation to push against that opacity. In fact, as these examples show, at its best the scientific community has worked closely with diverse communities to build broad coalitions of researchers and nonresearchers to assess and respond to risks. History offers both hope that building such collective processes are possible and repeated notes of caution about the difficulties of sustaining such necessary work. Three principles that emerge from across these historical case studies should inform how the scientific community leads present-day AI governance.

1. *Self-policing is not enough*: Researchers' voluntary moratoriums on publication or on specific research practices has rarely (if ever) proven sufficient, especially once impactful technologies have moved beyond controlled laboratory settings. Scientists and

engineers have been particularly poorly equipped to anticipate the ways in which public narratives about technologies would shape expectations and uses.

2. *Oversight must extend beyond the research community*: Broad-gauge input and oversight has repeatedly proven necessary to sustain an innovation ecosystem. Extended debate and negotiation among researchers and broader groups of nonspecialists can build public trust and establish clear regulatory frameworks, within which research can expand across academic and private-sector spaces.

3. *Recurring reviews are necessary*: In-depth reviews, conducted by reviewers that include specialists and broader communities of concerned stakeholders, should regularly reassess both the evolving technologies and the shifting social practices within which they are embedded. Only then can best practices be identified and refined. These reviews are most effective when they build on existing civic infrastructures and civil rights.

In all three historical examples, scientists and engineers were eager to act justly and to put bounds around novel technologies to mitigate potential risks. Yet these experts could not anticipate the ways in which official or popular enthusiasm would lead these innovations to spread in unexpected ways. For example, researchers did not predict the rise of an elaborate Cold War national security secrecy infrastructure, the reactions from Cambridge residents to fears of accidents or leaks involving dangerous pathogens, or the popular enthusiasm (despite legal skepticism) for the polygraph. These off-label uses, far beyond the reach of laboratory controls or formal legal strictures, have posed particular dangers to broader communities.

Nonetheless, by speaking decisively about risks, articulating clear gaps in knowledge, and identifying faulty claims, scientists and technologists—working closely with colleagues beyond the research community—have successfully established regulatory and governance frameworks within which new technologies have been developed, evaluated, and improved. The same commitment to genuine partnerships beyond the research community must guide governance of exciting—yet risky—AI technologies today.

Notes

1. Ruth Lewin Sime, *Lise Meitner: A Life in Physics* (Berkeley: University of California Press, 1996), chap. 10.

2. Niels Bohr and John A. Wheeler, "The Mechanism of Nuclear Fission," *Physical Review* 56 (September 1, 1939): 426–450. See also Sime, *Lise Meitner*, chap. 11.

3. Mark Walker, *German National Socialism and the Quest for Nuclear Power* (New York: Cambridge University Press, 1989), 17–18; Walter E. Grunden, Mark Walker, and Masakatsu Yamazki, "Wartime Nuclear Weapons Research in Germany and Japan," *Osiris* 20 (2005): 107–130; Margaret Gowing, *Britain and Atomic Energy, 1939–1945* (London: Macmillan, 1964), chap. 1; Richard G. Hewlett and Oscar E. Anderson, Jr., *A History of the United States Atomic Energy Commission*, vol. 1, *A New World, 1939–1946* (University Park: Pennsylvania State University Press, 1962), 15–17; David Holloway, *Stalin and the Bomb: The Soviet Union and Atomic Energy, 1939–1956* (New Haven: Yale University Press, 1994), chap. 3.

4. Alex Wellerstein, *Restricted Data: The History of Nuclear Secrecy in the United States* (Chicago: University of Chicago Press, 2021), chap. 1.

5. Alexei Kojevnikov, *Stalin's Great Science: The Times and Adventures of Soviet Physicists* (London: Imperial College Press, 2004), 132.

6. Wellerstein, *Restricted Data*, chap. 2; Hewlett and Anderson, *A New World*, chaps. 3–6.

7. Wellerstein, *Restricted Data*, chap. 4; Brian Balogh, *Chain Reaction: Expert Debate and Public Participation in American Commercial Nuclear Power, 1945–1975* (New York: Cambridge University Press, 1991), chaps. 4–5.

8. Lawrence Badash, *Scientists and the Development of Nuclear Weapons* (Atlantic City: Humanities Press, 1995), chaps. 5–6; Hewlett and Anderson, *A New World*, chaps. 12–16.

9. Kai Bird and Martin Sherwin, *American Prometheus: The Triumph and Tragedy of J. Robert Oppenheimer* (New York: Vintage, 2005), part 4; Jessica Wang, *American Science in an Age of Anxiety: Scientists, Anticommunism, and the Cold War* (Chapel Hill: University of North Carolina Press, 1999), chap. 1; David Kaiser and Benjamin Wilson, "American Scientists as Public Citizens: 70 Years of the *Bulletin of the Atomic Scientists*," *Bulletin of the Atomic Scientists* 71 (January 2015): 13–25.

10. Wang, *American Science in an Age of Anxiety*, chaps. 2, 5.

11. Bird and Sherwin, *American Prometheus*, part 5; Priscilla J. McMillan, *The Ruin of J. Robert Oppenheimer and the Birth of the Modern Arms Race* (New York: Viking, 2005), part 4; David Kaiser, "The Atomic Secret in Red Hands? American Suspicions of Theoretical Physicists During the Early Cold War," *Representations* 90 (Spring 2005): 28–60.

12. Michael Gordin, *Red Cloud at Dawn: Truman, Stalin, and the End of the Atomic Monopoly* (New York: Farrar, Straus, and Giroux, 2009); Francis J. Gavin, *Nuclear Statecraft: History and Strategy in America's Atomic Age* (Ithaca: Cornell University Press, 2012).

13. See also Peter Galison, "Removing Knowledge," *Critical Inquiry* 31 (Autumn 2004): 229–243.

14. See especially Sheldon Krimsky, *Genetic Alchemy: The Social History of the Recombinant DNA Controversy* (Cambridge, MA: MIT Press, 1982); and Susan Wright, *Molecular Politics: Developing American and British Regulatory Policy for Genetic Engineering, 1972–1982* (Chicago: University of Chicago Press, 1994).

15. Charles Weiner, "Drawing the Line in Genetic Engineering: Self-Regulation and Public Participation," *Perspectives in Biology and Medicine* 44 (Spring 2001): 208–220; John Durant, "'Refrain from Using the Alphabet': How Community Outreach Catalyzed the Life Sciences at MIT," in David Kaiser, ed., *Becoming MIT: Moments of Decision* (Cambridge, MA: MIT Press, 2010), 145–163.

16. David Baltimore, unpublished lecture in MIT Technology Studies Workshop, November 6, 1974, as quoted in John Durant, "'Refrain from Using the Alphabet,'" 146.

17. Paul Berg, David Baltimore, Herbert W. Boyer, Stanley N. Cohen, et al., "Potential Biohazards of Recombinant DNA Molecules," *Science* 185 (July 26, 1974): 303.

18. The impacts of the restriction of participation at Asilomar to life scientists is emphasized in Shobita Parthasarathy, "Governance Lessons for CRISPR/Cas9 from the Missed Opportunities at Asilomar," *Ethics in Biology, Engineering & Medicine: An International Journal* 6, nos. 3–4 (2015): 305–312; and J. Benjamin Hurlbut, "Remembering the Future: Science, Law, and the Legacy of Asilomar," in Sheila Jasanoff and Sang-Hyun Kim, eds., *Dreamscapes of Modernity: Sociotechnical Imaginaries and the Fabrication of Power* (Chicago: University of Chicago Press, 2015): 126–151.

19. Wright, *Molecular Politics*, chap. 4; Durant, "'Refrain from Using the Alphabet,'" 150.

20. Weiner, "Drawing the Line in Genetic Engineering"; Durant, "'Refrain from Using the Alphabet'"; David Kaiser and Jonathan D. Moreno, "Self-Censorship Is Not Enough," *Nature* 492 (20 December 2012): 345–347.

21. Mayor Alfred Vellucci, "Hearing on Recombinant DNA Experimentation, City of Cambridge," June 23, 1976, as quoted in Durant, "'Refrain from Using the Alphabet,'" 150.

22. Weiner, "Drawing the Line in Genetic Engineering"; Durant, "'Refrain from Using the Alphabet.'"

23. Durant, "'Refrain from Using the Alphabet,'" 150–156; see also Wright, *Molecular Politics*, 222.

24. Durant, "'Refrain from Using the Alphabet,'" 156–160. In other ways, the Cambridge-area biotech boom highlights enduring inequalities in the distribution and access to the advances of biotechnology. See Robin Scheffler, *Genetown: The Greater Boston Area and the Rise of Biotechnology in America* (Chicago: University of Chicago Press, forthcoming).

25. James H. Jones, *Bad Blood: The Tuskegee Syphilis Experiment*, rev. ed. (New York: Free Press, 1993); Eileen Welsome, *The Plutonium Files: America's Secret Medical Experiments in the Cold War* (New York: Dial 1999); Susan M. Reverby, "Ethical Failures and History Lessons: The US Public Health Service Research Studies in Tuskegee and Guatemala," *Public Health Reviews* 34, no. 1 (2012): article 13.

26. Laura Stark, *Behind Closed Doors: IRBs and the Making of Ethical Research* (Chicago: University of Chicago Press, 2012).

27. Stark, *Behind Closed Doors.*

28. Richard van Noorden, "The Ethical Questions That Haunt Facial-Recognition Research," *Nature* 587 (November 19, 2020): 354–358; Casey Fiesler, Nathan Beard, and Brian C. Keegan, "No Robots, Spiders, or Scrapers: Legal and Ethical Regulation of Data Collection Methods in Social Media Terms of Service," *Proceedings of the International AAAI Conference on Web and Social Media* 14, no. 1 (2020): 187–196.

29. Paul Ohm, "Broken Promises of Privacy: Responding to the Surprising Failure of Anonymization," *UCLA Law Review* 57 (2010): 1701–77; Jacob Metcalf and Kate Crawford,

"Where Are Human Subjects in Big Data Research? The Emerging Ethics Divide," *Big Data & Society* (June 2016): 1–14; Mary L. Gray, "Big Data, Ethical Futures," *Anthropology News*, January 13, 2017; Laura Stark, "Protections for Human Subjects in Research: Old Models, New Needs?," *MIT Case Studies in Social and Ethical Responsibilities of Computing*, no. 3 (Winter 2022); Simson Garfinkel, "Differential Privacy and the 2020 US Census," *MIT Case Studies in Social and Ethical Responsibilities of Computing*, no. 3 (Winter 2022).

30. Jimena Canales, *A Tenth of a Second: A History* (Chicago: University of Chicago Press, 2009), chap. 3.

31. Ken Alder, *The Lie Detector: The History of An American Obsession* (New York: Free Press, 2007).

32. Alder, *The Lie Detector*.

33. On "charismatic technologies," see Morgan Ames, *The Charisma Machine: The Life, Death, and Legacy of One Laptop per Child* (Cambridge, MA: MIT Press, 2019).

34. Alder, *The Lie Detector*.

35. Sheila Jasanoff, "Science on the Witness Stand," *Issues in Science and Technology* 6, no. 1 (Fall 1989): 80–87; David E. Bernstein, "Frye, Frye, Again: The Past, Present, and Future of the General Acceptance Test," *Jurimetrics* 41, no. 3 (2001): 385–408.

36. Simon A. Cole, "Toward Evidence-Based Evidence: Supporting Forensic Knowledge Claims in the Post-Daubert Era," *Tulsa Law Review* 43, no. 2 (2007): 263–84.

37. Chief Justice of the US Supreme Court, *Amendments to the Federal Rules of Evidence*, 118th Cong., 1st Sess., House Document 118-33, https://www.govinfo.gov/content/pkg/CDOC-118hdoc33/pdf/CDOC-118hdoc33.pdf. For additional discussion of the new rules, see, for example, https://www.law.cornell.edu/rules/fre/rule_702.

38. Office of Technology Assessment, *Scientific Validity of Polygraph Testing: A Research Review and Evaluation* (Washington, DC: Government Printing Office, 1983), 4.

39. Office of Technology Assessment, *Scientific Validity of Polygraph Testing*, 25.

40. Committee to Review the Scientific Evidence on the Polygraph, *The Polygraph and Lie Detection* (Washington, DC: National Academies Press, 2003).

41. See, for example, Andre A. Moenssens and Stephen B. Meagher, "Fingerprints and the Law," US Department of Justice, Office of Justice Programs, 2011, https://www.ojp.gov/library/publications/fingerprint-sourcebook-chapter-13-fingerprints-and-law.

42. Robert B. Stacey, "Report on the Erroneous Fingerprint Individualization in the Madrid Train Bombing Case," *Forensic Science Communications* 7, no. 1 (January 2005); National Research Council of the National Academies, *Strengthening Forensic Science in the United States: A Path Forward* (Washington, DC: National Academies Press, 2009); President's Council of Advisors on Science and Technology, *Forensic Science in Criminal Courts: Ensuring Scientific Validity of Feature-Comparison Methods* (Washington, DC: Executive Office of the President, September 2016).

43. Committee on Facial Recognition, *Facial Recognition Technology: Current Capabilities, Future Prospects, and Governance* (Washington, DC: National Academies Press, 2024).

44. Patrick Grother, Mei Ngan, and Kayee Hanaoka, "Face Recognition Vendor Test (FRVT), Part 3: Demographic Effects," Report NISTIR 8280 (Washington, DC: National Institute of Standards and Technology, December 2019); Sidney Perkowitz, "The Bias in the Machine," *MIT Case Studies in Social and Ethical Responsibilities of Computing*, no. 1 (Winter 2021).

45. Joy Buolamwini and Timnit Gebru, "Gender Shades: Intersectional Accuracy Disparities in Commercial Gender Classification," *Proceedings of the 1st Conference on Fairness, Accountability and Transparency* 81 (2018): 77–91, https://proceedings.mlr.press/v81/buolamwini18a.html; Martins Bruveris, Jochem Gietema, Pouria Mortazavian, and Mohan Mahadevan, "Reducing Geographic Performance Differentials in Face Recognition," *arXiv*, February 27, 2020, https://arxiv.org/abs/2002.12093; Philipp Terhörst, Mai Ly Tran, Naser Damer, Florian Kirchbuchner, et al. "Comparison-Level Mitigation of Ethnic Bias in Face Recognition," *Proceedings of the IEEE International Workshop on Biometrics and Forensics (IWBF)* (April 2020): 1–6.

46. Clare Garvie, Alvaro Bedoya, and Jonathan Frankel, *The Perpetual Lineup: Unregulated Police Face Recognition in America* (Washington, DC: Georgetown Law Center on Privacy and Technology, 2016).

47. See, for example, the literature review in Pawel Drozdowski, Christian Rathget, Antitza Dantcheva, Naser Damer, et al., "Demographic Bias in Biometrics: A Survey on an Emerging Challenge," *IEEE Transactions on Technology and Society* 1, no. 2 (June 2020): 89–103.

48. See, for example, Robert Williams, "I Was Wrongfully Arrested Because of Facial Recognition. Why Are Police Allowed to Use It?," *Washington Post*, June 24, 2020; Kashmir Hill, "Wrongfully Accused by an Algorithm," *New York Times* (June 24, 2020, updated August 3, 2020); Elaisha Stokes, "Wrongful Arrest Exposes Racial Bias in Facial Recognition Technology," *CBS News*, November 19, 2020; Kashmir Hill, "Another Arrest, and Jail Time, due to a Bad Facial Recognition Match," *New York Times*, December 29, 2020, updated January 6, 2021; Editorial Board, "Unregulated Facial Recognition Must Stop Before More Black Men Are Wrongfully Arrested," *Washington Post*, December 31, 2020.

49. Andrew G. Ferguson, *The Rise of Big Data Policing: Surveillance, Race, and the Future of Law Enforcement* (New York: NYU Press, 2017); Ruha Benjamin, *Race After Technology: Abolitionist Tools for the New Jim Code* (Medford, MA: Polity, 2019); Brian Jefferson, *Digitize and Punish: Racial Criminalization in the Digital Age* (Minneapolis: University of Minnesota Press, 2020).

50. Elly Cosgrove, "One Billion Surveillance Cameras Will Be Watching Around the World in 2021, a New Study Says," *CNBC*, December 6, 2019; M. Melton, "Government Watchdog Questions FBI on Its 640-Million-Photo Facial Recognition Database," *Forbes*, June 4, 2019; Perkowitz, "The Bias in the Machine."

51. Yilun Wang and Michal Kosinski, "Deep Neural Networks Are More Accurate than Humans at Detecting Sexual Orientation from Facial Images," *Journal of Personality and Social Psychology* 114, no. 2 (2018): 246–57, https://doi.org/10.1037/pspa0000098; cf. Jacob Metcalf, "'The Study Has Been Approved by the IRB': Gayface AI, Research Hype, and the Pervasive Data Ethics Gap," *Medium*, November 30, 2017, https://medium.com/pervade-team/the-study-has-been-approved-by-the-irb-gayface-ai-research-hype-and-the-pervasive-data-ethics-ed76171b882c. For a sample of news coverage of the Wang and Kosinski study, see, for example, Alan Burdick, "The A.I. 'Gaydar' Study and the Real Dangers of Big Data," *New Yorker*, September 15, 2017; Heather Murphy, "Why Stanford Researchers Tried to Create a 'Gaydar' Machine," *New York Times*, October 9, 2017; Brian Resnick, "This Psychologist's 'Gaydar' Research Makes Us Uncomfortable; That's the Point," *Vox*, January 29, 2018; and Paul Lewis, "'I Was Shocked It Was So Easy': Meet the Professor Who Says Facial Recognition Can Tell If You're Gay," *The Guardian*, July 7, 2018.

52. van Noorden, "Ethical Questions," 354–358; Fiesler, Beard, and Keegan, "No robots, spiders, or scrapers," 187–196.

53. Kashmir Hill, *Your Face Belongs to Us* (New York: Penguin Random House, 2023). The trend is not limited to the United States. See, for example, Kai Strittmatter, *We Have Been Harmonised: Life in China's Surveillance State* (London: Old Street, 2019); and Tristan G. Brown, Alexander Statman, and Celine Sui, "Public Debate on Facial Recognition Technologies in China," *MIT Case Studies in Social and Ethical Responsibilities of Computing*, no. 2 (Summer 2021).

54. Jenna Burrell, "How the Machine 'Thinks': Understanding Opacity in Machine Learning Algorithms," *Big Data & Society* 3, no. 1 (June 1, 2016); see also Ho Chit Siu, Kevin J. Leahy, and Makai Mann, "STL: Surprisingly Tricky Logic (for System Validation)," *2023 IEEE/RSJ International Conference on Intelligent Robots and Systems (IROS)* (2023): 8613–8620, https://ieeexplore.ieee.org/document/10342290.

CHAPTER 5

Navigating AI Governance as a Normative Field: Norms, Patterns, and Dynamics

Urs Gasser

Introduction

This article conceptualizes AI governance as an emerging normative field by offering a series of analytical lenses and a set of initial observations aimed at contributing toward a navigation aid for what promises to be a rapidly evolving and complex ecosystem. The main objective of this contribution is to make visible the broad range of approaches, strategies, and instruments available in the governance toolbox as decision-makers in the public and private sectors seek to anticipate, analyze, and address harms and risks associated with the accelerating pace of AI development, deployment, and use while harnessing its potential for humans, society, and the planet at large.

This article is written at a moment in time when a myriad of AI governance initiatives are underway at the national, regional, and global levels, involving a broad range of actors, incentives, and interests. Such efforts range from comprehensive legislative projects like the EU AI Act[1] and whole-of-government efforts like the US Executive Order on Safe, Secure, and Trustworthy AI[2] and its accompanying implementation initiatives, to voluntary commitments and best practice frameworks. They include local governance interventions at the city level and international initiatives put forward by organizations like the Council of Europe,[3] the United Nations,[4] or G7[5] and G20,[6] to name just a few examples. Other important components of evolving AI governance arrangements include ethical as well as technical standards, developed again across all levels of governance, ranging from company-level to international-level.

Taken together, AI governance as a "hot field" (borrowing a term coined by sociologist Robert Merton) consists of a heterogeneous set of principles, norms, rules, standards, and decision-making procedures. In governance parlance, it fits within the broader concepts of multilevel, multiactor, and multimodal governance, despite recent trends toward an enhanced role of governments as regulators.[7] At least at present and for the foreseeable future, AI governance can be understood as a case of polycentric governance, to invoke a concept developed by Elinor Ostrom,[8] with multiple centers of decision-making and overlapping responsibilities, without a single entity that has the ultimate authority for making all collective decisions.

Given the polycentric nature and fluid state of AI governance, this contribution does not aim to describe or evaluate any single effort in greater depth or to arrive at policy recommendations. Rather, it seeks to offer a series of lenses through which contemporary initiatives can be analyzed and contextualized. Such a descriptive approach might inform future normative frameworks by offering a sense of various approaches and instruments available and by highlighting some of the factors shaping their application.

The first section frames AI governance as a normative field and situates it within the broader context of ever-evolving technology as a socially embedded venture shaped by numerous factors and forces at play. The following section, "Approaches to AI Governance," offers several elements of a possible taxonomy of approaches to AI governance that shape the contours and interactions among a diverse set of principles, norms, rules, standards, and decision-making procedures. It suggests a number of lenses that might be useful when understanding and navigating the range of options available to steer the development, deployment, and use of AI. Embracing the complexity and heterogeneity of AI governance as a normative field, the subsequent section, "Mapping Normative Patterns," seeks to identify a series of normative patterns within and across different AI governance arrangements, with a focus on recent legal developments. The last sections of the chapter aim to demarcate conceptual zones of convergence, divergence, and possible interoperability across different AI governance arrangements

("Selected Nodes of AI Governance"), and to offer final considerations for AI governance-making as shaping the further evolution of a normative field ("AI Governance for an Uncertain Future").

AI Governance as a Normative Field

This section frames AI governance as a normative field, starting with a working definition, followed by a brief overview of some of the most salient initiatives and building blocks of AI governance arrangements both nationally and internationally. By briefly contextualizing AI governance in a broader social context, it also offers a reminder that neither the technology nor efforts to govern it have emerged in a vacuum.

Defining AI Governance

Defining the contours of AI governance is not an easy task.[9] The definition of what accounts for AI has been contested all along and varies across contexts and actors. Despite various efforts, a uniform standard definition has not emerged yet—and even some of the most influential definitions are subject to updates, as the recent definitional amendments to AI Guidelines of the Organisation for Economic Co-operation and Development (OECD) illustrate.[10] The challenge of defining where AI governance starts and where it ends is further exacerbated not only because the term *AI* is contested but also because the notion of governance is a highly amorphous concept with many meanings across different cultural and application contexts.[11] Questions of terminology seem mostly of academic interest at first, and it is striking that languages generally have received relatively little attention in contemporary AI conversations. But when entering the regulatory arena, more precise understandings of certain terms matter greatly and have real-world consequences, as the struggles to specify the many newly introduced terms in the EU AI Act might illustrate.

This article avoids a sharp definition of the subject it seeks to explore and takes a pragmatic approach. With respect to AI, the updated definition by the OECD serves as the term's core with a halo around it, reflecting broader definitions used in other norm complexes aimed at steering the development, deployment, and use of AI across a spectrum of open and closed technological and organizational settings. Similarly, a pragmatic understanding of the concept of governance is adopted, embracing a diversity of modalities of norms (from ethical principles to hard law), different levels of governance (from local to global), and a range of actors involved in such efforts (from professional associations to lawmakers).

Taking these elements together, AI governance can be circumscribed as the sum of all coexisting forms of collective regulation of matters associated with machine-based systems, which infer from inputs how to generate outputs that have the potential to influence physical or virtual environments.

Emerging AI Governance Arrangements

Fueled by an accelerating pace of innovation in AI research, development, and deployment, debates about the needs for and modalities of AI governance have intensified in recent years, spanning local to global levels. A broad range of stakeholders has launched various initiatives to set up dedicated guardrails for AI-based technologies, starting with several hundred AI ethics principles initiatives,[12] followed by hundreds of legislative and regulatory interventions,[13] as well as a plethora of standard-setting and best practice efforts. Among the many initiatives, the following flagship efforts with the potential for international impact serve as reference points in this chapter:

- *Canada's Draft Artificial Intelligence and Data Act*[14] introduces guardrails to ensure that AI systems deployed in Canada are safe and nondiscriminatory and creates accountability mechanisms for businesses as they develop and use AI-based technologies.
- *China's Interim Generative AI Measures*[15] seek to encourage and guide the responsible use of generative AI with respect for national

security while making everyone who develops and uses generative AI products to provide services to the public in China subject to government oversight.

- *Brazil's Draft Artificial Intelligence Act*[16] seeks to create rules for making AI systems available in Brazil, establish rights for people affected by their operation, provide penalties for violations, and set up a supervising body.
- *EU's AI Act*[17] is a comprehensive draft law aimed at addressing the risks of AI through a broad range of obligations and requirements to safeguard the health, safety, and fundamental rights of citizens. It seeks to ensure the proper functioning of the EU single market by setting consistent rules for AI systems across the EU.
- *US Executive Order on Safe, Secure, and Trustworthy AI*[18] establishes new and whole-of-government standards for government agencies to address safety and security risks associated with the development and use of AI in the social, economic, and national security spheres.

These initiatives only provide a subset of the diverse AI governance arrangements at the national level. The US AI governance landscape, for instance, consists of an amalgam of norms, which includes—in addition to the Executive Order and bills such as the US Algorithmic Accountability Act[19]—sector-specific initiatives (e.g., in the health and transportation sectors) and legislation at the state and city level, in addition to a broad range of soft law instruments ranging from an AI Bill of Rights[20] to voluntary commitments by leading AI companies, numerous ethical principles by private and public sector entities, and technical standards by standard-setting organizations such as the National Institute of Standards and Technology (NIST),[21] to name just a few AI governance sources.

Other nation states—including the United Kingdom, India, Japan, Singapore, and Switzerland—have taken a different route so far (note that things remain in flux) by either pursuing a sectoral approach to AI governance or refraining from the use of hard law while promoting the responsible development, deployment, and use of AI through non-binding governance mechanisms such as guidelines, best practices, and standards.

While some of these efforts at the national and regional level target AI specifically as a distinct set of technologies using different techniques and methods, AI governance has not emerged in isolation. Existing *general* guardrail regimes, among other factors discussed subsequently, provide the relevant normative context in which more *specific* interventions now take place.

Contextualizing AI Governance

AI governance, like AI itself, should not be considered in isolation but rather contextualized as part of a social fabric of norms and stabilized expectations, ranging from formalized policies and laws to often more implicit cultural values and attitudes.[22] They shape and limit what is possible, feasible, and desirable within a given ecosystem when addressing the broad range of opportunities and challenges associated with AI through means of governance.

Approaches to AI governance arrangements are situated within broader economic, social, environmental, technology, and regulatory *policies* of countries. Within these general parameters, many nations have enacted *national AI strategies*, which often also outline the contours of the envisioned AI regime.[23] A comparative analysis of AI strategies across twenty-two countries suggests a typology of prescribed governance approaches, resulting in a matrix with strong versus weak state interventions on one axis and stimulation versus enclosure-and-control approaches on the other. Different roles of the state in AI governance can be mapped onto each resulting quadrant, indicating certain levels of activity and the use of preferred governance instruments.[24]

Preexisting laws are another contextual element, as briefly mentioned. Consider, for instance, how relatively relaxed privacy laws or safe harbor provisions have contributed to an AI innovation-friendly ecosystem in the

United States.[25] Conversely, other sets of norms have arguably constrained some of the conditions conducive to AI advancement. While the empirical effects of stricter data protection laws in Europe on the development and adoption of AI remain contested, some studies suggest that the General Data Protection Regulation (GDPR) and particularly more stringent enforcement actions shaped important dimensions of the research and innovation ecosystems.[26]

The relevant context of AI governance is of course not limited to policy and law. Powerful forces that shape the present and future of AI governance originate from the spheres of economic and national security interests—an important nexus that goes beyond the scope of this chapter.[27] For context, it suffices to acknowledge that the shapes both of general legal norms and specific AI guardrails are heavily influenced by the *political economy*, understood as the actions taken by different stakeholders with divergent interests and unequal resources and power that characterize a given environment.[28] The extensive lobbying efforts by large technology companies, for instance, to push for guardrails that are favorable to their businesses are well-known and have also become apparent in the AI context. Perhaps more than anything, *geopolitical dynamics*—both in terms of competition and cooperation—frame the broader normative picture in which AI governance activities unfold in each domain and region,[29] and have led to what some have described as a "race to AI regulation" on top of the global race for AI.[30] The AI policy of the European Union, for instance, was positioned from the outset against the backdrop of global developments,[31] and its AI Act has already been analyzed through the prism of the so-called Brussels effect.[32]

These and several other factors—including *culturally anchored values, preferences, and attitudes* by people toward innovative technologies[33]—influence the normative context in which present day AI governance efforts crystallize. In other words, emerging AI governance norms are not endogenous rules but are socially embedded. The AI policies of Nordic nations, for instance, distinctly rely on core cultural values as organizing principles to steer the development of AI in society.[34] These normative dynamics complicate any comparison between different regimes and, above all, limit the possibility of successfully transplanting legal and other AI norms from one context to another.

International Initiatives

The AI governance landscape at the national and regional level is also shaped by a series of important international developments and initiatives,[35] including the influential *OECD Principles on Artificial Intelligence*,[36] which seek to promote AI that is innovative and trustworthy and that respects human rights and democratic values; the *UNESCO Recommendations on the Ethics of AI*[37] that spans standard-setting, policy advice, and capacity building; the UK-led *Bletchley Declaration*[38] that concerns international coordination on frontier AI; the *G7 Hiroshima AI Process* that promotes guardrails for advanced AI systems at the global level, among several others, including efforts such as the formation of a *UN High-Level Advisory Body on AI* and, more recently, the *UN Resolution on Safe, Secure and Trustworthy Artificial Intelligence System*.[39]

As this incomplete list already indicates, international efforts also range from relatively high-level aspirational principles to binding instruments. With respect to the latter, the most important initiative is the Council of Europe's (CoE) *Framework Convention on Artificial Intelligence Human Rights, Democracy, and the Rule of Law*.[40] The treaty covers the use of AI systems in both public and private sectors (with notable exceptions in areas such as national security), offering two compliance pathways when regulating the private sector: direct obligation to the treaty's provisions or alternative measures while respecting international human rights, democracy, and the rule of law. This accommodates global legal diversity. It mandates transparency, oversight, risk assessment, and mitigation measures, including identifying AI-generated content and assessing the need for moratoriums or bans on high-risk AI uses.

The treaty ensures AI systems uphold equality, privacy rights, and accountability for adverse impacts, with legal remedies for human rights violations and procedural safeguards. It requires parties to adopt measures to ensure that AI systems do not undermine democratic institutions and establishes a Conference of the Parties for follow-up, and it requires independent oversight to ensure compliance, raise awareness, and foster public debate on AI technology.[41]

Relevant building blocks of international AI governance that predate some of the most recent global AI initiatives can also be found in the domain of *free trade and digital economy agreements.* For instance, the Digital Economy Partnership Agreement between Singapore, Chile, and New Zealand, promoting interoperability among the different digital trade regimens, promoted the adoption of ethical AI frameworks and developed mechanisms for cross-border data flows.[42] The UK-New Zealand Free Trade Agreement, to take another example, removed certain data localization requirements and established guardrails for international data flows between the two countries.[43]

Institutionalized initiatives also include *regional and bilateral efforts.*[44] Under the institutional umbrella of the US-EU Trade and Technology Council (TCC), for instance, the United States and the European Union committed to a series of projects to advance trustworthy AI through collaborations in the area of measurement and evaluation, the design of AI tools to protect privacy, and the economic analysis of AI's impact on workforce. An initial contribution is the TCC Joint Roadmap on Evaluation and Measurements Tools for Trustworthy AI and Risk Management, with commitments to work toward a common terminology (a draft of an EU-US Terminology and Taxonomy for Artificial Intelligence was recently released for consultation) and a common knowledge base of metrics and methodologies to coordinate their work with international standard bodies, and track emerging risks and work toward compatible evaluations of AI systems. Progress has also been made in the area of privacy and AI workforce impact analysis.

Cross-Pollination

Even below the threshold of larger institutionalized international efforts, and despite the previously mentioned idiosyncrasies that point toward nuanced AI governance arrangements across geographies and contexts, the process of developing such arrangements at the local and national levels is currently characterized by a remarkable degree of *cross-pollination* among policymakers and lawmakers.[45] Put differently, not only do geopolitical dynamics shape the normative field of AI governance, but the approaches and instruments that are deployed within the respective spheres of polycentric governance-making are themselves shaped by interactions among relevant stakeholders, elevating the complexity of the norm dynamics at play.

Forums and venues where such processes take place range from informal Zoom calls, conferences, and workshops to engagement in committees and networks, such as the Global Partnership on AI Governance or the G20 Working Group on Artificial Intelligence, to name just a few examples specific to the domain of AI. Platforms such as Globalpolicy.AI, the Transatlantic Policy Network, and the World Economic Forum, or collaborations between think tanks such as the Brookings Institution and the Center for European Policy Studies, also serve as important spaces for cross-pollination among various stakeholders, including policymakers and lawmakers, in addition to direct lines of communication among them. (Members of the US Congress, for instance, have engaged with one of the rapporteurs of the EU AI Act.) Efforts facilitated by academic institutions, such as the Stanford Institute for Human-Centered AI, also serve as exchange points for decision-makers in the field of AI.

Cross-pollination through knowledge diffusion in the field of AI governance takes place though various other mechanisms with varying degrees of informality and transparency. Examples include structured interactions in the context of standards-setting organizations involved in AI governance—the collaboration between the OECD and NIST to develop a catalog of AI tools and metrics is a case in point—but also lobbying efforts by industry and industry associations that often operate across jurisdictions and promote certain approaches or instruments across different forums of AI governance-making.

Toward Governance Innovation?

Each cycle of technological innovation with the potential to induce structural shifts in a socioeconomic environment when interacting with humans and society typically challenges existing governance structures. While the default response to such challenges is to apply the old structures to the new phenomenon, the disruptions also offer a window of opportunity for *innovation within governance systems*. Some of these governance innovations are gradual in nature and others more radical; some include novel institutions, and others innovate around processes or rights.[46] The internet revolution, for instance, led to several governance innovations across all three domains, with ICANN being an example of an institutional innovation, online dispute resolution systems a process innovation, and the right to be forgotten a rights innovation.[47]

While traces of innovative governance might be spotted at the levels of individual norms within large governance projects such as the EU AI Act, it is the calls for *new AI oversight institutions* voiced by government representatives, industry leaders, and academics that have recently garnered public attention. The new models proposed for AI governance often find inspiration in other policy domains, including climate, finance, or nuclear energy. A recent review of proposals for new AI institutions clustered them into seven functional categories that transcend traditional government policies. Models range from scientific and political consensus-building to coordinating institutions in the realm of policy and regulation, and from enforcement of standards and restrictions to international joint research and distribution of benefits and access to AI technology.[48]

The analysis suggests a wide array of models and experiences that can be leveraged as the quest for global AI governance intensifies. In the current quicksilver environment, it is arguably one of the most intriguing and consequential questions how much innovation in governance is needed and (politically) possible to unlock the benefits of AI while managing its risk at the global level, and what would such an arrangement look like in practice.[49]

Approaches to AI Governance

When returning to present day approaches to AI governance, the complexity and heterogeneity of the evolving AI governance landscape, the contextually embedded nature of the respective normative arrangements, and the speed of development make it difficult to meaningfully engage in a comparative norm-level analysis between and among different initiatives across various levels of governance. What this section offers, instead, are a number of analytical lenses that can be used to help understand and position different governance approaches relative to each other, highlighting the broad range of conceptual and functional pathways available.

Positioning Approaches

With these caveats in mind, one might take a closer look at the diverse AI governance arrangements that together form the normative field. Given the number of initiatives and the fluid state of norm development around the world, it is virtually impossible within the scope and purpose of this article to offer even a representative, let alone a comprehensive overview of current attempts aimed at governing AI. A more modest approach is to position some of the most salient governance initiatives along several spectrums with ideal-type approaches at their respective ends:

- *Sectoral versus horizontal approaches*: AI-based technologies cover various application contexts. Governance approaches can seek to regulate AI horizontally across their different use cases or regulate the development, deployment, and use for specific sectors, such as health, transportation, justice, or education, to name just a few. The United Kingdom takes a sectoral approach; other country examples include Japan and Switzerland. At least traditionally, the United States has also pursued a sectoral approach, with the recent Executive Order blurring the lines to the extent it pursues a whole-of-government approach. The EU, with its EU AI Act and related efforts like the AI Liability Directive, takes a decidedly horizontal approach to AI governance, supplemented by sector-specific regulations, resulting in a mixed approach, but partly also interacting with other legislation, including the GDPR but also the Digital Services Act (DSA)—the latter in which foundation models are incorporated in very large online platforms and search engines.[50]

- *Soft law versus hard law*: Another positioning point is the question whether a given AI governance approach relies more on soft law or hard law. Soft law instruments include standards, ethics guidelines, checklists, best practices, to name a few. They play a key role in self-regulatory regimes, but in the field of AI, they also supplement state-driven legislation and regulation. Japan and Singapore currently rely heavily on soft law instruments, which continue to play a prominent role in the United States, for instance, in the gestalt of the Voluntary Commitments[51] from leading companies to manage the risks posed by AI, but also the NIST AI Risk Management Framework,[52] among others. Ambitious hard law approaches are currently pursued in the European Union, Canada, and Brazil, but also part of the AI governance mix in the United Kingdom (sectoral regulation) and the United States (particularly state and local levels), as already mentioned.

The spectrums outlined here interact with each other and partially overlap. As already indicated, AI governance initiatives often combine different approaches and instruments within them. For instance, hard law approaches to AI governance will typically also rely on standard-setting outside the formal lawmaking processes, as subsequently discussed. While not being exclusive and clear-cut, the spectrums might still serve as a rough coordination system to identify the position of different approaches relative to each other.

Cutting across the sectoral versus horizontal and soft versus hard law approaches are two other spectrums that can be helpful when considering the available toolbox of AI governance and comparing different strategic choices made by AI governance bodies:

- *Outcomes versus procedural approaches*: Outcome-based approaches to AI governance stipulate a desirable outcome such as innovation, economic growth, or safety, to name just a few objectives, and keep the means to achieve these objectives typically flexible. Procedural approaches, in contrast, prescribe instruments that need to be adopted along the way, assuming that they will lead to a desirable outcome. Risk management is a case in point. Risk-based approaches categorize AI applications based on their level of risk to individuals and society and attach tailored requirements to each level. Perhaps the most prominent example in the latter category is the EU AI Act, with its intricate scheme of risk classification and corresponding legal obligations. The Canadian draft legislation also builds upon a risk-based approach, as well as the Brazilian Draft AI Law, which the EU AI Act inspired. Examples of the former approach include the United Kingdom's pro-innovation approach to AI governance, which targets the outcomes AI will likely generate in specific applications rather than assigning rules according to risk levels.

- *Principles versus rules-based approaches*: As the name suggests, principle-based approaches seek to guide the development, deployment, and use of AI by laying down a set of overarching principles that guide the relevant stakeholders. Prominent examples of such an approach are the OECD AI Principles and the G7 International Guiding Principles on AI. Rules-based approaches, in contrast, typically lay out specific and more detailed rules according to which the relevant stakeholders must play. At the country level, China's Interim Measures in the realm of generative AI are illustrative. But also sector-specific requirements, for instance, in the area of medical AI or transportation, might often be rule-based, suggesting again that approaches might be mixed and are often not clear-cut.

Across these four spectrums (and others could be added), it is important to remember that any categorization of this sort runs the risk of oversimplification and is of limited value given the characteristics of AI as a "messy" normative field as discussed in the earlier section. At the very least, however, they might illustrate the range of approaches available and serve as a rough navigation aid when contrasting different choices made by various AI governance actors.

Functional Dimensions

Another lens for positioning approaches to AI governance—and within such approaches, individual norms—is functional in nature. Borrowing from analyses of previous cycles of technological innovation and accompanying governance responses in law and regulation, one can distinguish between constraining, enabling, and leveling functions of norms.[53]

Lawmakers or regulators might draft and enact norms that *constrain* the development or use of certain types of technologies or functionalities. Frequently used instruments include the following:

- *Prohibitions*: Legislation or regulation can ban the development or use of certain AI systems or applications. The EU AI Act, for instance, prohibits certain AI use cases that pose unacceptable risks. Its far-reaching restrictions on using facial recognition technology are another case in point. Similarly, several US municipalities have restricted or banned the use of facial recognition technology by local agencies. Restrictions on the export of dual-use technologies are another illustration. More generally, the Canadian AIDA proposes new criminal law provisions to prohibit any reckless and malicious uses of AI that cause serious harm to Canadians and their interests.
- *Premarket obligations*: AI laws and regulations can stipulate requirements that need to be met before an AI product enters the market. The EU AI Act requires developers of high-risk AI systems to perform comprehensive conformity assessments before placing them on the market. In areas such as medical AI or autonomous vehicles premarket approval is often required, including under US regulations. *Post*-market monitoring is another instrument in the toolbox, often supplementing premarket regulatory schemes, like in the case of high-risk systems under the EU AI Act.
- *Certification and registration*: Analytically distinct, but closely related to premarket requirements are certification and registration schemes. The important role of (at least) voluntary certifications is mentioned in the in the Canadian Artificial Intelligence and Data Act (AIDA), for example. High-Risk AI systems under the

proposed EU AI Act are subject to a strict certification regime. In addition, such systems, as well as foundational models, need to be registered in an EU database.

Enabling norms, in contrast, are designed to permit or even promote the development and use of technology. Such norms—most prominently so-called safe harbors—have played an important role in creating a flourishing digital platform economy, as mentioned before. In the AI realm, laws and regulations can promote the development and use of AI systems in various ways, ranging from compliance exceptions for certain use cases to government investments. Some of the common instruments include the following:

- *Funding and subsides:* Enabling AI legislation can establish funding schemes to support the development or adoption of AI-based technologies. Lawmakers across countries and regions, including the United States and Europe, have enacted laws as a foundation to direct investment and subsidies toward industry as well as the public sector. The US Executive Order, for instance, directs federal funding toward a research coordination network to support privacy preserving technologies, among several other actions.
- *Capacity building:* AI legislation might stipulate capacity building measures. Instruments may range from technical assistance programs to setting up resource and innovation centers. The US Executive Order, for instance, provides small developers and entrepreneurs with access to technical assistance and resources. The envisioned AI and Data commissioner under the Canadian AIDA is also engaged in capacity building.
- *Sandboxes:* Various AI laws encourage or even mandate regulatory sandboxes, which allow businesses and regulators to cooperate in a controlled environment to test innovative products or services and gain insights with regard to risks of these innovation and appropriate safeguards. Sandboxes are a key measure in support of innovation under the EU AI Act, for instance. The Brazilian AI Act set the foundation for testing environments to support the development of innovative AI systems.

Norms that are *leveling* the playing field include, for instance, general rules prohibiting anti-competitive behavior or deceptive business practices. Specific instruments that seek to address information asymmetries or other imbalances in the AI context include the following:

- *Transparency*: To bridge information gaps, AI laws and regulations often impose disclosure obligations, which come in many forms and shapes. The EU AI Act, for instance, requires clear and comprehensible information about the capabilities and limitations of high-risk AI systems, and transparent and traceable decision-making processes. The Brazilian AI Act, to take another example, mandates transparency in the use of AI systems in interactions with natural persons, among other requirements.
- *Education and training*: AI literacy and skill-building programs might also have their anchor in laws and regulations. The US Executive Order, for instance, supports various programs to enhance AI-relevant skills to ensure access to AI opportunities for the workforce in general, and for specialized groups of professionals, such as investigators or prosecutors.

To be sure, this list of instruments is not exhaustive; additional mechanisms currently under consideration cover a broad spectrum of governance techniques, including licensing requirements,[54] tax obligations, rulemaking authority, and procurement power, among others. Furthermore, several additional instruments transcend the three categories of constraining, enabling, and leveling norms. For instance, auditing and inspection regimes, oversight mechanisms, or sanctions are frequently used techniques to create accountability and ensure compliance and thus serve a *cross-cutting function*.

Mixed Approaches

Clearly, the evolving normative field of AI governance is complex and the traditional taxonomies start to blur. Particularly at the country level, AI

governance often involves mixed approaches, combining different strategies and instruments and situating these countries somewhere along the
spectrum of the ideal-type approaches outlined earlier. Moreover, although the dimensions that mark each spectrum might be analytically
distinct, they are also interacting. The EU approach to AI governance is a
helpful illustration in this regard: The EU AI Act advances a risk-based
approach through hard law but is supplemented by sectoral regulations in
areas such as health and transportation as well as soft law instruments
such as technical standards and ethical principles. As already indicated,
the coordination system is perhaps most helpful to understand the relative
positioning between different approaches and to create awareness of the
available options, particularly for countries and communities that remain
undecided on which approach to pursue.

The same applies to the functional categories. AI governance, like previous tech-induced governance regimes, typically consists of a complex
amalgam of norms. Such arrangements typically combine several of the
functions briefly described earlier, as a recent empirical study of several
hundred of proposed (and at times enacted) AI laws and regulations across
the Atlantic indicates.[55] Nonetheless, certain trends become visible when
applying a functional lens. Mapping proposed, rejected, and enacted legislation on AI-based technologies in the United States and Europe over the
past seven years, the study reveals that legislative activities on both sides of
the Atlantic serve different functions, but (proposed) laws and regulations
in the United States tend to be more strongly in the enabling zone than their
European counterparts.

Effectiveness and Ripple Effects

Mapping existing approaches to AI governance in general, and a high-level
overview of some of the most salient instruments that are available to lawmakers and regulators in particular, indicates a *deep reservoir of normative
techniques* (both social and technological in nature) that AI governing bodies can tap into when seeking to steer the development, deployment, and
use of AI across diverse application areas. While the choices among

approaches and instruments are not unconstrained and, as discussed in the subsection "Contextualizing AI Governance", shaped by numerous factors that create path-dependencies, the respective actors involved often have significant leeway when selecting and mixing the tools to address specific AI governance issues.

AI governance shares characteristics of a wicked policy problem with many interdependencies and contingencies, making it virtually impossible to predict in all nuances the individual and aggregated effects when choosing one governance approach over another, or when selecting certain instruments while not deploying others.[56] Experiences from previous cycles of technology innovation offer some high-level insights for the design of "good governance" and clues about possible ramifications of different approaches at a basic level.[57] For instance, comparisons between US and European approaches to privacy and data protection, or an analysis of different governance regimes across regions when regulating online intermediaries such as social media platforms, might teach some lessons.[58] However, at the more granular level of specific instruments, lawmakers and regulators often fly in the dark, as links between interventions and desirable outcomes (for instance, in terms of effectiveness, efficiency, and flexibility when addressing a given AI issue) remain chronically uncertain when dealing with structural sociotechnological transitions.[59]

In the normative field of AI governance, as in other domains, a complex web of economic, social, technological, organizational, and also human factors influences the practical outcomes emerging from any given mix of governance approaches and instruments over time. Constraining norms as a subset of AI governance arrangements and the question of pressures and incentives that might affect compliance and enforceability are indicative of some of the complicating dynamics. Research on the effects of different approaches to privacy and (pre-GDPR) data protection in the United States and European countries, for instance, revealed that *on-the-ground practices*—including overall awareness, leadership buy-in, and professional culture—have been critical factors determining privacy outcomes regardless of the underlying conceptual choices made by lawmakers and regulators.[60] Another example is a finding from a recent in-depth examination of China's hard law approach to AI governance concerning generative AI, suggesting a significant gap between the "law on the books" and "*law in action*" when it comes to the willingness to enforce the strict rules amid the geopolitical arms race and vis-à-vis domestic economic struggles.[61] At a more abstract level, both stories—albeit for different reasons—point toward the importance of communities of practice and implementation capacities, respectively, that in no small part will co-determine the effectiveness of any of the available approaches to AI governance.

The uncertainties and dynamics involved when dealing with emerging science and technology such as AI make it not only challenging to select and combine approaches in ways that best address a given governance issue but also very difficult to anticipate *second order and ripple effects*. Governance instruments that promote the use of AI in the public administration, for instance, might exacerbate environmental issues or have implications for AI supply chains. AI guardrails that do not stand the test of time might affect public trust not only in the technology but also in the state. For all these reasons, it is vital to incorporate performance benchmarks and evaluation processes as well as mechanisms of responsible experimentation and systematic learning into AI governance arrangements, whether based on soft or hard law or on a sectoral or horizontal approach.[62]

Mapping Normative Patterns

Analyzing contemporary AI governance arrangements as building blocks of emerging governance regimes is a difficult task, as the caveats in the previous sections already indicate. AI governance as a normative field, as mentioned before, is not yet defined by clear boundaries; rather it is a moving target where general background rules and specific norms enacted by a broad range of governance actors interact. Given the polycentric character of current AI governance arrangements, various norm types are involved, with varying degrees of abstraction, levels of legitimacy, and prescriptive power. Finally, AI governance norms emerge within specific institutional, legal, and cultural contexts, but they also interact with each other as discussed earlier, making comparisons among them challenging.

One promising methodological approach to deal with this complexity and heterogeneity is to look out for *normative patterns* instead of comparing individual norms. The approach, which is inspired by the theory of law as normative patterns in a normative field as developed by the late Swedish legal theorist Anna Christensen (who in turn got inspiration from Douglas Hofstadter's analysis of AI programs), is based on the empirical observation that different basic normative patterns can be distinguished both within and across a multitude of norms that seek to collectively regulate social matters.[63]

Applying this idea of normative pattern analysis to the normative field of AI governance, several patterns emerge when looking within and across the EU AI Act and US Executive Order in particular, but also when considering selected other laws such as the Brazilian and Canadian AI bills, as well as soft laws and international governance initiatives.[64]

Protection of Established Rights

Various legal norms in evolving AI governing arrangements seek to ensure and bolster the protection of established rights of rightsholders vis-à-vis novel risks and potential harms associated with the development and use of AI. Together, this cluster of norms forms one of the key patterns that transcend the heterogeneous set of norms of AI governance. Within the EU AI Act, for instance, the protection of established rights plays an important role and goes to the heart of the raison d'être of the legislation, which is set out to ensure a high level of protection of health, safety, fundamental rights, democracy, the rule of law, and the environment from harmful effects of AI systems. Similarly, several sections of the US Executive Order aim to protect established rights, for instance, when stipulating requirements against unlawful discrimination, protection against fraud, and threats to privacy, or to ensure the safety, security, and reliability of AI systems. Brazil's proposed AI legislation, for example, included a section on the protection of the rights of individuals impacted by AI decision-making and outlined individual and collective rights of action. Rights protection is also a core motif of soft law instruments advanced globally. Consider, for example, Singapore's Model AI Governance Framework, which includes the protection of the interests of human beings, including their well-being and safety, as primary considerations in the design, development, and deployment of AI. Also, in the international realm, various AI governance initiatives include explicit or implicit references to protecting established rights. The G7 Hiroshima Process International Code of Conduct for Organizations Developing Advanced AI Systems with its requirement to respect human rights and protect children and vulnerable groups, or the Bletchley Declaration with its recognition that the protection of human rights, safety, privacy, and data protection needs to be addressed, or the CoE's Framework Convention with its rules to align the life cycle of AI systems with international and national legal protections of human rights, are examples of pars pro toto.

Protection of Established Positions

A second, related pattern that crystallizes across a diverse set of AI governance arrangements at the national and international levels is the protection of established positions, where some of the legal norms aimed at governing AI are crafted in ways that are protective of previously recognized economic, cultural, and social interests and aimed at preserving a given status quo. The EU AI Act includes various normative references along these lines. At the systemic and most fundamental level, measures taken to shield democracy and justice, for instance by limiting certain uses of AI or imposing strict upfront requirements, are examples of the protection of established positions. Regarding the protection of individual interests, the EU AI Act clarifies that it does not alter any of the previous rules, particularly in the realm of data protection, consumer protection, and product safety, which establish important baselines in terms of protected interests and positions. In other cases, it seeks to reaffirm established interests, for instance, with respect to copyright holders' economic interests in the context of the regulation of foundation models used in generative AI systems, where providers would be required to publicly disclose a sufficiently detailed summary of the copyrighted material used as training data. In the same way, the Brazilian AI Law reinforces previous regulations, especially those related to data and consumer protection. Concerning intellectual property, it establishes that it is without prejudice to the owners of these rights. US Executive Order also includes norms aimed at protecting established positions, for instance, in the form of requirements aimed at improving the security, resilience, and incident response related to AI usage in critical infrastructure or when outlining different measures and programs in support of workers who might face future AI-related job disruptions, including the protection of their economic interests and well-being. Protecting established positions is also a key driver behind international AI governance initiatives. The Bletchley Declaration, to take just one example, with its focus on measures to ensure the safety of AI systems, for instance, in the context of frontier AI capabilities, is motivated to safeguard protected interests and existing positions of individuals, organizations, and governments.

Market Functional Pattern

The market functional pattern is a dynamic element in AI governance arrangements. Norms at the core of this pattern aim to promote new economic activities, stimulate technological development through market mechanisms, and support new markets and business models. Improving the smooth functioning of the internal market while promoting the uptake of human-centric and trustworthy AI and ensuring high levels of protection is among the overarching objectives of the EU AI Act. It enables AI systems, with notable exceptions, to benefit from the principle of free movement of goods and services. References to market functioning are spread across the proposed law and mentioned in the context of open software and data as enablers of market-based research and innovation. Transparency requirements are also contextualized as minimally invasive measures to avoid unjustifiable restrictions on trade. More generally, the EU AI Act is embedded in a broader digital strategy aimed at enhancing Europe's competitiveness and promoting innovation in the digital market. On the other side of the Atlantic, the US Executive Order also states as a core principle the promotion of responsible innovation and competition and stresses the importance of a fair, open, and competitive ecosystem and marketplace for AI and related technologies. Its requirements to promote innovation and competition, but also to nurture AI talent and strengthen US leadership internationally are, to a large extent, part of a market functional pattern. At least traces of the same pattern can also be found in soft law instruments. The Singaporean Model AI Framework,[65] for instance, contextualizes its best practices in terms of AI as an enabler of new goods and services and a booster of productivity and competitiveness, which can lead to economic growth and better quality of life. At the international level, the market functional pattern has been less explicit in recent AI governance initiatives, with occasional references to productivity gains and inclusive economic growth, for instance, in the Bletchley Declaration. Market functional rationales also pop up in various AI-related efforts, such as in the realm of data governance aimed at enabling trans-border flows of data, building upon the international order of IP and trade as the normative bedrock of globalized markets.

Fostering Innovation

As already mentioned in the context of the functional dimensions of AI governance, several AI governance arrangements contain dedicated norms to promote research and development of AI-based technologies. Expanding on the original concept of normative pattern analysis pioneered by Anna Christensen, this complex of norms can be conceptualized as the fostering of innovation pattern. It interacts with the market functional pattern and is often framed as the protection of the potential for innovation based on preexisting commitments to free trade and intellectual property. Again, this normative pattern is typically present in national-level AI governance arrangements and in international initiatives and cuts across the hard law and soft law distinction. The provisions of the EU AI Act mention innovation close to thirty times. It stipulates various norms aimed at promoting innovation or protecting the potential for innovation, ranging from the possibility of regulatory sandboxes and coordinated standard-setting in the technical realm to AI literacy initiatives, among others. Likewise, the proposed Brazilian legislation also adopts regulatory sandboxes to promote innovation. Norms aimed at promoting AI innovation are also integral to the UK White Paper, which proportionately tailors its regulatory framework to fulfill the goal of innovation promotion,[66] and to the US Executive Order, which outlines a broad range of measures to promote innovation and competition through immigration reform, investments in resources, support for research and development, and measures in the realm of IP protection, spanning various governmental agencies and bolstering private-public partnerships. Soft law instruments of AI governance often also include recommendations aimed at promoting innovation, both at the national and international levels. The influential OECD AI Principles, for example, call on governments to consider long-term public investments and encourage private investments in research and development to spur innovation in trustworthy AI, including in creating open datasets to support the overall environment for responsible AI research. Along similar lines, at the global level, the UNESCO Recommendation on the Ethics of AI calls upon member states to ensure that public funds are dedicated to responsible and inclusive AI research and that governments promote international collaboration to advance innovation.

The patterns proposed in this chapter, inspired by Christensen's original work, are an attempt at describing some of the core normative elements within and across different AI governance arrangements. Given the complexity and heterogeneity of AI governance arrangement, these different patterns do not make up a hierarchy of norms; rather they coexist and interact with each other in ways shaped by various contextual factors, including cultural, political, and economic conditions as alluded to in the previous section when contextualizing AI governance. Although the mode of analysis is descriptive rather than prescriptive, the approach can serve as a foundation to study how patterns manifest themselves over time within and across different societal conditions and application contexts.

Selected Nodes of AI Governance

The previous sections sketched some of the approaches, functions, and patterns of AI governance as a moving normative field, highlighting by example the great variety of pathways available when seeking to regulate the development, deployment, and use of AI. Building on this mapping exercise, this section looks at cross-cutting crystallization points in some of the AI governance arrangements featured in this article and intends to highlight both zones of convergence and divergence in the normative field. Again, several AI governance initiatives at the national and international levels are referenced to illustrate some of the commonalities and differences among them at the conceptual level. Last, the section suggests and identifies early traces of an interoperability approach as a potential way forward to navigate both zones of convergence and divergence across AI governance arrangements.

Zones of Convergence

While much nuance remains, some trends of convergence can be observed across most of the AI governance arrangements reviewed in this chapter. For the methodological reasons mentioned before, the following commonalities focus on conceptual "nodes" of AI governance rather than on individual norm-level comparisons.

- *Prominence of risk-based approaches*: While some AI governance actors opt for outcome-based approaches, risk-based approaches to AI governance have gained popularity at both the national and international levels, cutting across sectoral and horizontal as well as soft and hard law instruments. Leading examples at the national level include the EU AI Act, Canadian AI and Data Act, and Brazilian AI bill, which all use some forms of risk and impact assessment to group AI systems into different categories of compliance obligations. The US Executive Order also highlights the importance of a risk-based approach, particularly when managing risks from the federal government's own use of AI and in the context of implementation measures, for instance, in the gestalt of the NIST AI Risk Management Framework as an influential voluntary standard. Other soft law instruments, such as the Singaporean Model AI Governance Framework, provide guidance to organizations to adopt a risk-based approach when implementing measures. At the international level, risk-based approaches have been promoted by G7 digital and technology ministers and referenced in the Hiroshima Process International Code of Conduct for Organizations Developing Advanced AI Systems.[67] Although these examples suggest conceptual convergence toward risk-based approaches, substantial differences continue to exist at the operational level.[68]

- *Role of regulatory sandboxes*: Building on previous experiences using sandboxes as a supervised experimental space to enable responsible testing of emerging technologies and foster bidirectional learning between developers and regulators, AI governance bodies across the globe have started to embrace this technique and are currently applying it to AI. The European Union in the EU AI Act and several European member states, including Spain and Germany, are promoting the use of AI regulatory sandboxes as controlled environments with reduced regulatory burden to keep pace with rapid AI development while gaining experience dealing with it effectively. The Brazilian AI Act also authorizes the operation of an experimental regulatory environment for innovation in AI, and the preparation for a first sandbox is already underway. Singapore, as a last example, recently launched a Generative AI Evaluation Sandbox as an experimental platform for developers to build responsible AI use cases and enable the evaluation of trusted AI products.[69]

- *Importance of standards*: Across all the reviewed AI governance arrangements, regardless of their respective positioning, standards play a vital role.[70] Even in cases where comprehensive legislation is at the core of AI governance, like in Europe with the EU AI Act, standard-setting is a critical part of the strategy. The European Committee for Standardization (CEN) and the European Committee for Electrotechnical Standardization (CENELEC) are leading organizations developing standards that could provide developers the presumption of conformity with the EU AI Act. As already mentioned, NIST in the United States has been actively involved in developing standards for AI by creating a framework that fosters the development of trustworthy and responsible AI systems, covering areas such as bias, explainability, and robustness. The International Organization for Standardization (ISO) and the International Electrotechnical Commission (IEC) have a joint technical committee focused on AI standardization, and international organizations like the Institute of Electrical and Electronics Engineers (IEEE) have set up various working groups developing standards for ethical considerations in AI, to name just a few initiatives among many.

Zones of Divergence

The complexity and heterogeneity of the AI governance landscape make it unsurprising that many differences exist not only at the level of individual norms—for instance, whether a given AI governance arrangement specifically addresses foundation models, and if so, how—but also at the conceptual level. In addition to the higher-level differences resulting from distinct approaches to AI governance already mentioned earlier in "Approaches to AI Governance," some of the particularly noteworthy conceptual areas of divergence include the following:

- *Scope and definitions*: Many important nuances exist regarding the scope of application among the variety of different governance initiatives, as well as with respect to definitions of various technical and legal terms. Voluntary industry standards and professional best practices, for instance, have a very different scope and reach than mandatory laws and regulations, whether horizontal by design or sector-specific. The EU AI Act, for instance, seeks to regulate the full range of AI applications across the private and public sectors, whereas the aforementioned US Algorithmic Accountability

Act appears to be more specific and selective. The US Executive Order, in contrast, takes a broad whole-of-government approach to AI governance. While progress has been made when it comes to developing a shared understanding of key terms such as AI itself at the international level—strongly influenced by the important work of the OECD, which recently updated the definition of AI—many other important definitions of key concepts are a work-in-progress or remain contested, as recent debates about the definition of all-purpose AI, generative AI, and foundation models in the EU AI Act illustrate. To be sure, differences in terminology are neither a new phenomenon nor unique to AI governance. However, in light of the heterogeneous norms landscape, it remains a significant challenge for the years to come to create appropriate levels of (semantic) interoperability across a thickening web of emerging laws, standards, and best practices that might apply simultaneously given the polycentric nature of AI governance.

- *Level of normative commitment*: Despite the flourishing of AI governance initiatives in general and recent momentum around the creation of hard laws after a phase with a strong emphasis on ethical norms, stark differences remain among such efforts when it comes to the level of the underlying normative commitment. Perhaps most significantly and visibly, the commitment to individual rights varies greatly across AI governance arrangements. This applies not only when comparing AI norms of environments governed by the rule of law versus others but also when considering the depth of normative guarantees offered by different democratic regimes. For instance, while the US Executive Order marks without any doubt an important step forward when it comes to the protection of civil rights and privacy against emerging AI risks, it does not immediately offer the same level of actionable legal protection for individuals as the provisions set forth in the EU AI Act and the General Data Protection Regulation, respectively. Similarly, most, if not all, the AI governance requirements stipulated in soft laws such as Singapore's AI Model AI Governance Framework as well as many of the global AI governance initiatives, are important signals and milestones on a longer trajectory but still are relatively weak normative commitments when assessed from the vantage point of advancing individual rights beyond the current baseline of human rights protections.

- *Enforcement*: The AI governance arrangements reviewed in the context of this chapter (and beyond) vary greatly in terms of enforcement regimes. As a threshold, much depends initially on the specifics of the governance approach itself, for instance, the role of voluntary self-regulation versus government-based regulation through hard law. Consistent with the polycentric characteristics of the AI governance landscape, different norms are typically enforced by different actors, ranging from in-house AI accountability boards or professional associations to traditional law enforcement or from newly created AI authorities to preexisting specialized agencies tasked with AI norm enforcement in their respective sectors or industries. How an AI enforcement regime looks is yet again shaped by broader contextual factors, including preexisting legal order and market structure. For instance, the EU AI Act puts the primary enforcement responsibility in the hands of the member states, with consultation and coordination mechanisms at the EU level in the form of a European Artificial Intelligence Board chaired by the EU Commission. At least from a structural perspective, the EU approach to enforcement resembles regimes of harmonized data protection law, including strong enforcement tools. Adopting the same strategy, the Brazilian AI Law establishes a new supervisory authority that will be responsible for monitoring noncompliance with the law, promoting its implementation, and issuing other regulations related to AI. In the United States, the Federal Trade Commission (FTC)—in addition to the agencies now tasked with the implementation of the US Executive Order—is expected to play a particularly important role as an enforcer of consumer protection–oriented AI norms and standards at the federal level (complemented by specialized agencies such as the FDA in health AI), but without a comprehensive mandate compared to the EU counterparts.

Zones of Interoperability

The concept of interoperability offers an alternative analytical view on the diverse landscape of emerging AI governance arrangements—a perspective that transcends the binary division between zones of convergence and divergence. Originally a technical concept, interoperability in the digital realm can be broadly understood as the ability of different systems, applications, or components to work together based on the exchange of useful data and other information.[71] Under the header of "legal interop," it has been analogized to conceptualize the working together among distinct legal norms across jurisdiction that regulate the global flow of information.[72] A number of instruments are available to enhance legal interoperability, including legal harmonization, mutual recognition, reciprocity, cooperation, and standardization—approaches that can be operationalized through various means, ranging from treaty law to self-regulation.[73] Some of these tools might also be relevant when seeking to enhance interop between AI governance arrangements or their components.

- Many of the *international initiatives* led by state and nonstate actors mentioned earlier in this chapter are often aimed at enhancing the interoperability of norms, rules, standards, and decision-making procedures across different AI governance arrangements. The OECD Principles on Artificial Intelligence or the UNESCO Recommendations on the Ethics of AI, for instance, have informed and often shaped hard law and soft law approaches across various jurisdictions, promoting interoperability at the norm and process levels and beyond. The G7 Hiroshima Process also aims to establish common guiding principles for organizations developing advanced AI systems while acknowledging that "different jurisdictions may take their own unique approaches to implementing these guiding principles in different way."[74] More recent efforts such as the United Nations Resolution encourage "internationally interoperable identification, classification, evaluation, testing, prevention and mitigation of vulnerabilities and risks" of AI systems.[75]

- Higher levels of interop, however, might not only come from top-
  down efforts. *Multistakeholder initiatives* can also enable the working together among different arrangements and regimes.[76] In the
  field of AI governance, such efforts are still in the early stages, but
  important work is well underway.[77] The Global Partnership on
  Artificial Intelligence, for instance, has produced various guides
  on the responsible development, use, and adoption of AI.[78] The
  Partnership on AI, too, has advanced best practices in various
  areas of AI governance, including synthetic media.[79] The AI
  Governance Alliance, convened by the World Economic Forum,
  produced interoperable building blocks to guide the safe development, deployment and use of generative AI across AI governance
  arrangements.[80] ETH Zurich in collaboration with the Swiss government hosts a multistakeholder Gen AI Redteaming network to
  collaborate on disclosing, replicating, and mitigating safety issues
  and develop best practices.[81]

As discussed, emerging AI governance arrangements introduce and le-
gitimize a variety of innovative approaches, tools, and practices—ranging
from human rights and risk assessments to codes of practices—that will
need to be further specified and operationalized in different forums and
processes. From an interop perspective, this *modularization* of AI governance opens the possibility for cross-border multistakeholder cooperation,
with the promise to enhance alignment between different AI governance
arrangements by enabling the *working together* among some of their core
components even absent more ambitious harmonization at the international
regime level.[82]

AI Governance for an Uncertain Future

This chapter has explored AI governance as a normative field from a pre-
dominantly descriptive perspective. Developing detailed prescriptions at
the level of concrete norms from such an initial mapping exercise during the
early stage of AI governance with little empirical evidence about what works

under what conditions is at least problematic from a methodological perspective. Therefore, the analytical lenses introduced in the previous section and the discussion of possible normative patterns within and across AI governance arrangements suggest at least a number of considerations when contemplating additional interventions to regulate the development, deployment, and use of AI. Specifically, the discussion in this chapter offers five key takeaway points.

First, the complexity and heterogeneity of AI governance as an evolving normative field suggest the adoption of an ecosystem perspective when considering additional initiatives aimed at steering the development, deployment, and use of AI. Metaphorically speaking, the AI governance landscape resembles more a tropical garden rather than a formal garden with neatly trimmed lawns, arranged flower beds, and precise geometric designs. Without pushing the analogy too far, future AI governance interventions like tropical gardening require interaction with the sociotechnological environment, a deep understanding of the cultural, societal, economic, legal, and other relevant contexts, and a sense for integrating governance initiatives within the surrounding environment.

Second, the mapping of various AI governance arrangements along a number of interacting spectrums—such as sectoral versus horizontal approaches, soft versus hard law, outcomes versus risk-based, or principles versus rules-based approaches—as well as the different functions of governance norms, principles, standards, and decision-making procedures point toward a broad range of available approaches, strategies, and tools in the AI governance toolkit. Future regulatory initiatives should consider the full range of instruments available and select them based on their fit for purpose when addressing specific AI governance issues. Ultimately, the selection of tools will need to be guided not only by features such as efficacy and efficiency but also by overarching values such as legitimacy, accountability, and fairness.

Third, any future governance initiative needs to be designed and implemented with context in mind. The discussion in the preceding sections has highlighted a number of such contextual factors and alluded to legal path-dependencies, the political economy, and geopolitical dynamics among the forces at play. While AI governance arrangements from other

Table 5.1. How functions, instruments, and normative patterns interact within and across AI governance arrangement and what (selected) governance issues they typically address

Functions	Instruments and mechanisms	Normative patterns	Main AI governance issues	Examples
Constraining	Prohibitions	Protection of established rights; protection of established positions; market functional patterns	Existential risk; democratic erosion; freedom and autonomy	Chapter II EU AI Act (Prohibited AI Practices)
	Pre-market obligations		Performance outcomes, incl. security, safety, privacy, nondiscrimination (bias), fairness	Chapter III Section 2 EU AI Act (Requirements for high-risk AI systems)
	Certification, registration		Responsibility, incl. accountability	Chapter III Section 5 EU AI Act (. . . conformity assessment, certificates, registration)
Enabling	Funding, subsidies	Fostering innovation; protection of established positions	Performance outcomes; sustainability; geopolitical competition	US Executive Order (various provisions)
	Capacity building	Fostering innovation; market functional patterns	Performance outcomes; Geopolitical competition	US Executive Order (various provisions)
	Sandboxes		Performance outcomes, evidence-based policy	Art. 38 and Art. 39 Brazilian AI Act Draft (measures to encourage innovation)

Table 5.1. (*Continued*)

Functions	Instruments and mechanisms	Normative patterns	Main AI governance issues	Examples
Leveling	Transparency	Market functional patterns; protection of established positions; protection of established rights	Explainability; trustworthiness; accountability	S. 11 Canadian AIDA Draft (publication of description)
	Education, training	Fostering innovation; market functional patterns	Labor displacement; job quality; performance outcomes	US Executive Order (various provisions)
Cross-cutting	Rulemaking	Protection of established rights; protection of established positions; market functional patterns	Accountability, compliance, enforcement	US Executive Order (various provisions)
	Auditing			US Executive Order (various provisions)
	Oversight			S. 33 Canadian AIDA Draft (AI and data commissioner)
	Sanctions			Chapter XII EU AI Act (incl. penalties)

Note: The table connects the sections "Approaches to AI Governance" and "Mapping Normative Patterns" and benefits from the concise overview of AI governance issues and interventions by the Working Group on Regulation and Executive Action of the National AI Advisory Committee (NAIAC), "Rationales, Mechanisms, and Challenges to Regulating AI: A Concise Guide and Explanation," Non-Decisional Statement.

contexts—for instance, from other regions—might serve as sources of inspiration, recent experiences with the General Data Protection Regulation offer a cautionary tale when it comes to legal transplants that ignore the contextual realities in which they are supposed to be adopted. Debates about a possible Brussels effect originating from the EU AI Act need to consider these complexities and limitations, particularly vis-à-vis majority world countries.

Fourth, the select initiatives touched on in this chapter, featuring a small subset of AI governance arrangements currently in the making, give a sense not only of the heterogeneity of relevant principles, norms, standards, and decision-making processes but also point toward an enormous degree of complexity at the implementation level. Future AI governance initiatives should not only specify what problem they seek to address in what context and through what means, but in parallel invest in capacity building to enable and empower key actors both in the private and public sectors to turn abstract principles and norms into actual practices. Such capacity building requires multistakeholder and increasingly international cooperation and has significant implications for the education and training of civil servants and private sector leaders alike.

Last, the descriptive engagement with selected elements of various AI governance arrangements suggests a series of broader design questions when it comes to guardrail-making amid an increasingly discontinuous future in front of us. From the vantage point of guardrail design more generally, AI governance—and not only AI—should be human-centric by guiding and supporting individuals to make better decisions considering socially desirable outcomes that define us as communities and hold us together as societies. Such a perspective not only suggests a critical examination of the suitable principles, norms, standards, and decision-making processes to govern AI but also highlights the importance of appropriate requirements that guide the design of such rules, including principles such as guardrail diversity, variability, plasticity, and self-constraint.[83]

Notes

Thanks to Martha Minow and Susan Ness as well as the participants of the NAS-Sunnylands-APPC AI Retreat for helpful comments on an earlier draft, and to Noha Lea Halim and Jiawei Zhang for research assistance. Manuscript as of March 24, 2024. Contact: Urs.Gasser@tum.de.

1. European Parliament, *Artificial Intelligence Act* (March 2024), https://www.europarl.europa.eu/doceo/document/TA-9-2024-0138_EN.pdf.

2. The White House, *Executive Order on the Safe, Secure, and Trustworthy Development and Use of Artificial Intelligence* (October 30, 2023), https://www.whitehouse.gov/briefing-room/presidential-actions/2023/10/30/executive-order-on-the-safe-secure-and-trustworthy-development-and-use-of-artificial-intelligence/.

3. Council of Europe, *Council of Europe Framework Convention on Artificial Intelligence and Human Rights, Democracy and the Rule of Law* (September 2, 2024), https://rm.coe.int/1680afae3c.

4. United Nations General Assembly, *Seizing the Opportunities of Safe, Secure and Trustworthy Artificial Intelligence Systems for Sustainable Development: Draft Resolution* (March 11, 2024), https://digitallibrary.un.org/record/4040897?v=pdf&ln=en.

5. Ministry of Foreign Affairs of Japan, *G7 Leaders' Statement on the Hiroshima AI Process* (October 30, 2023), https://www.mofa.go.jp/ecm/ec/page5e_000076.html.

6. Ministry of External Affairs, Government of India, *G20 New Delhi Leaders' Declaration* (September 9, 2023), https://www.mea.gov.in/bilateral-documents.htm?dtl/37084/G20_New_Delhi_Leaders_Declaration.

7. See, for example, Urs Gasser and Virgilio A. F. Almeida, "A Layered Model for AI Governance," *IEEE Internet Computing* 21, no. 6 (November 20, 2017), https://ieeexplore.ieee.org/document/8114684.

8. See, for example, Elinor Ostrom, *Understanding Institutional Diversity* (Princeton, NJ: Princeton University Press, 2005).

9. See also Araz Taeihagh, "Governance of Artificial Intelligence," *Policy and Society* 40, no. 2 (June 4, 2021), 137–157.

10. OECD, *Recommendation of the Council on Artificial Intelligence* (adopted May 21, 2019, amended May 2, 2024), https://legalinstruments.oecd.org/en/instruments/OECD-LEGAL-0449.

11. See, for example, Gunnar Folke Schuppert, *The World of Rules: A Somewhat Different Measurement of the World* (Frankfurt/Main, Germany: Max-Planck-Institut für Rechtsgeschichte und Rechtstheorie, 2017).

12. See, for example, Anna Jobin, Marcello Ienca, and Effy Vayena, "The Global Landscape of AI Ethics Guidelines," *Nature Machine Intelligence* 1 (September 2, 2019), https://doi.org/10.1038/s42256-019-0088-2.

13. See, for example, Nestor Maslej et al., "The Artificial Intelligence Index Report 2023," AI Index Steering Committee, Institute for Human-Centered AI, Stanford University (Stanford, CA: April 2023), https://aiindex.stanford.edu/wp-content/uploads/2023/04/HAI_AI-Index-Report_2023.pdf.

14. Canada, Parliament, House of Commons, *An Act to Enact the Consumer Privacy Protection Act, the Personal Information and Data Protection Tribunal Act and the Artificial Intelligence and Data Act and to Make Consequential and Related Amendments to Other Acts*, 1st sess., 44th Parliament, 2021, https://www.parl.ca/legisinfo/en/bill/44-1/c-27.

15. Cyberspace Administration of China (CAC), "Interim Measures for the Management of Generative Artificial Intelligence Services" [in Chinese] (July 10, 2023), https://www.cac.gov.cn/2023-07/13/c_1690898327029107.htm.

16. Brazilian Federal Senate, *Dispõe sobre o uso da Inteligência Artificial*, Bill No. 2338/2023 [in Portugese] (2023), https://www25.senado.leg.br/web/atividade/materias/-/materia/157233.

17. European Parliament, *Regulation (EU) 2024/1689 of the European Parliament and of the Council of 13 June 2024 Laying Down Harmonised Rules on Artificial Intelligence and Amending Regulations (EC) No 300/2008, (EU) No 167/2013, (EU) No 168/2013, (EU) 2018/858, (EU) 2018/1139 and (EU) 2019/2144 and Directives 2014/90/EU, (EU) 2016/797 and (EU) 2020/1828 (Artificial Intelligence Act) (Text with EEA Relevance)*, https://eur-lex.europa.eu/eli /reg/2024/1689/oj.

18. White House, *Executive Order on the Safe, Secure, and Trustworthy Development and Use of Artificial Intelligence.* See also Office of Management and Budget, "OMB Releases Implementation Guidance Following President Biden's Executive Order on Artificial Intelligence," The White House, November 1, 2023, https://www.whitehouse.gov/omb/briefing-room/2023 /11/01/omb-releases-implementation-guidance-following-president-bidens-executive -order-on-artificial-intelligence/.

19. Algorithmic Accountability Act of 2023, H. R. 5628 (2023).

20. Office of Science and Technology Policy, *Blueprint for an AI Bill of Rights*, The White House (2022), https://www.whitehouse.gov/wp-content/uploads/2022/10/Blueprint-for-an-AI -Bill-of-Rights.pdf.

21. National Institute of Standards and Technology, "AI Standards," August 3, 2021, updated June 5, 2024, https://www.nist.gov/artificial-intelligence/ai-standards.

22. See, for example, Susana Borrás and Jakob Edler, Eds., *The Governance of Socio- Technical Systems: Explaining Change* (Cheltenham: Edward Elgar, 2014).

23. See, for example, Laura Galindo, Karine Perset, and Francesca Sheeka, "An Overview of National AI Strategies and Policies," OECD Going Digital Toolkit, Policy Note, 2021, https://goingdigital.oecd.org/data/notes/No14_ToolkitNote_AIStrategies.pdf.

24. Christian Djeffal, Markus B. Siewert, and Stefan Wurster, "Role of the State and Responsibility in Governing Artificial Intelligence: A Comparative Analysis of AI Strategies," *Journal of European Public Policy* 29, no. 11 (2022): 1799–1821, https://doi.org/10.1080/13501763 .2022.2094987.

25. See, for example, Ajay Agrawal, Joshua Gans, and Avi Goldfarb, "Economic Policy for Artificial Intelligence," *Innovation Policy and the Economy* 19, no. 1 (2019), https://doi.org /10.1086/699935.

26. See, for example, Katherine Quezada-Tavarez, Lidia Dutkiewicz, and Noémie Krack, "Voicing Challenges: GDPR and AI Research," *Open Research Europe* 2:126 (November 23, 2022), https://open-research-europe.ec.europa.eu/articles/2-126; Nicholas Martin, Christian Matt, Crispin Niebel, and Knut Blind, "How Data Protection Regulation Affects Startup Innovation," *Information Systems Frontiers* (2019), https://doi.org/10.1007/s10796-019-09974 -2; for a more general discussion, see Panel for the Future of Science and Technology, *The Impact of the General Data Protection Regulation (GDPR) on Artificial Intelligence* (Brussels: European Union, 2020), https://www.europarl.europa.eu/RegData/etudes/STUD/2020 /641530/EPRS_STU(2020)641530_EN.pdf.

27. See, for example, Paul Scharre, *Four Battlegrounds: Power in the Age of Artificial Intelligence* (New York: W.W. Norton 2023).

28. See, for example, Maximilian Kasy, "The Political Economy of AI: Towards Democratic Control of the Means of Prediction," Institute for New Economic Thinking, the Oxford Martin School, April 14, 2023, https://oms-inet.files.svdcdn.com/production/files /handbook_politicalecon_ai.pdf.

29. See, for example, Inga Ulnicane et al., "Governance of Artificial Intelligence: Emerging International Trends and Policy Frames," in Maurizio Tinnirello, Ed., *The Global Politics of Artificial Intelligence* (New York: Chapman and Hall/CRC, 2022).

30. Nathalie A. Smuha, "From a 'Race to AI' to a 'Race to AI Regulation': Regulatory Competition for Artificial Intelligence," *Law, Innovation and Technology* 13, no. 1 (March 23, 2021), https://doi-org.proxy.library.upenn.edu/10.1080/17579961.2021.1898300.

31. See, for example, Alessandro Annoni et al., "Artificial Intelligence: A European Perspective," *European Union* (Luxembourg: Publication Office of the European Union, 2018), doi:10.2760/11251.

32. See Anu Bradford, *Digital Empires: The Global Battle to Regulate Technology* (Oxford: Oxford University Press, 2023).

33. See, for example, Sönke Ehret, "Public Preferences for Governing AI Technology: Comparative Evidence," *Journal of European Public Policy* 29, no. 11 (2022): 1779–1798, https://doi.org/10.1080/13501763.2022.2094988.

34. Stephen Cory Robinson, "Trust, Transparency, and Openness: How Inclusion of Cultural Values Shapes Nordic National Public Policy Strategies for Artificial Intelligence (AI)," *Technology and Society* 63 (2020), https://doi.org/10.1016/j.techsoc.2020.101421.

35. See, for example, Lewin Schmitt, "Mapping Global AI Governance: A Nascent Regime in a Fragmented Landscape," *AI and Ethics* 2 (August 17, 2022): 303–314.

36. Organisation for Economic Co-operation and Development, "OECD AI Principles Overview," oecd.ai, n.d., https://oecd.ai/en/ai-principles.

37. UNESCO, *Recommendation on the Ethics of Artificial Intelligence*, United Nations (2022), https://unesdoc.unesco.org/ark:/48223/pf0000381137.

38. AI Safety Summit, *The Bletchley Declaration* (November 1, 2023), https://www.gov.uk/government/publications/ai-safety-summit-2023-the-bletchley-declaration/the-bletchley-declaration-by-countries-attending-the-ai-safety-summit-1-2-november-2023.

39. Office of the Secretary-General's Envoy on Technology, "High-Level Advisory Body on Artificial Intelligence," United Nations (n.d.).

40. *Council of Europe Framework Convention on Artificial Intelligence and Human Rights, Democracy and the Rule of Law*, Council of Europe (2024), https://rm.coe.int/1680afae3c.

41. Council of Europe Newsroom, "Council of Europe Adopts First International Treaty on Artificial Intelligence," Council of Europe, May 17, 2024, https://www.coe.int/en/web/portal/-/council-of-europe-adopts-first-international-treaty-on-artificial-intelligence.

42. "Digital Economic Partnership Agreement (DEPA)," New Zealand Ministry of Foreign Affairs & Trade, n.d., https://www.mfat.govt.nz/en/trade/free-trade-agreements/free-trade-agreements-in-force/digital-economy-partnership-agreement-depa.

43. Department for Business and Trade and the Department for International Trade, "UK-New Zealand FTA: Data Explainer," gov.uk, February 28, 2022, https://www.gov.uk/government/publications/uk-new-zealand-fta-data-explainer.

44. For a detailed overview, see Cameron F. Kerry et al., "Strengthening International Cooperation on AI," Brookings/Center for European Policy Studies (October 25, 2021). For a framework, see Pekka Ala-Pietilä and Nathalie A. Smuha, "A Framework for Global Cooperation on Artificial Intelligence and its Governance," in B. Braunschweig and M. Ghallab, *Reflections on AI for Humanity (Preprint)* (New York, NY: Springer, 2021).

45. For a general overview of the barriers to cross-cultural cooperation on AI and how to overcome them, see, for example, Seán S. ÓhÉigeartaigh et al. "Overcoming Barriers to Cross-cultural Cooperation in AI Ethics and Governance," *Philosophy & Technology* 33 (2020): 571–593.

46. Urs Gasser and Herbert Burkert, "Regulating Technological Innovation: An Information and a Business Law Perspective" in *Rechtliche Rahmenbedingungen des Wirtschaftsstandortes Schweiz: Festschrift 25 Jahre juristische Abschlüsse an der Universität St. Gallen* (Zürich: Dike, 2007).

47. See, for example, Urs Gasser and John Palfrey, *Advanced Introduction to Digital Law* (Cheltenham: Edward Elgar, forthcoming 2025).

48. Matthjis Maas and José Jaime Villalobos, "International AI institutions: A Literature Review of Models, Examples, and Proposals," Legal Priorities Project (September 2023).

49. See Alondra Nelson, "The Right Way to Regulate AI: Focus on Its Possibilities, Not Its Perils," *Foreign Affairs*, January 12, 2024.

50. See, for example, Cornelia Kutterer, "Regulating Foundation Models in the AI Act: From 'High' to 'Systemic' Risk," AI Regulation Papers (January 2024).

51. The White House, "FACT Sheet: Biden-Harris Administration Secures Voluntary Commitments from Leading Artificial Intelligence Companies to Manage the Risks Posed by AI," The White House, September 12, 2023, https://www.whitehouse.gov/briefing-room/statements-releases/2023/07/21/fact-sheet-biden-harris-administration-secures-voluntary-commitments-from-leading-artificial-intelligence-companies-to-manage-the-risks-posed-by-ai/

52. "AI Risk Management Framework," NIST, 2024, https://www.nist.gov/itl/ai-risk-management-framework.

53. Gasser and Burkert, "Regulating Technological Innovation."

54. See, for example, Neel Guha et al., "The AI Regulatory Alignment Problem," HAI Policy & Society and Stanford RegLab (November 2023).

55. Kerstin N. Vokinger, David Schneider, and Urs Gasser, "Mapping Legislative and Regulatory Dynamics of Artificial Intelligence in the US and Europe" (September 2023, manuscript under review).

56. Tim Büthe et al., "Governing AI—Attempting to Herd Cats? Introduction to the Special Issue on the Governance of Artificial Intelligence," *Journal of European Public Policy* 29, no. 11 (2022): 1721–1752, https://doi.org/10.1080/13501763.2022.2126515.

57. See, for example, Inga Ulnicane et al., "Good Governance as a Response to Discontents? Déjà Vu, or Lessons for AI from Other Emerging Technologies," *Interdisciplinary Science Reviews* 46, no. 1–2 (March 7, 2021): 71–93, https://doi-org.proxy.library.upenn.edu/10.1080/03080188.2020.1840220.

58. See, for example, Urs Gasser and Wolfgang Schulz, "Governance of Online Intermediaries: Observations from a Series of National Case Studies," Berkman Center Research Publication Series No. 2015–5 (February 2015).

59. See, for example, Stefan Kuhlmann, Peter Stegmaier, and Kornelia Konrad, "The Tentative Governance of Emerging Science and Technology—A Conceptual Introduction," *Research Policy* 48 (2019), 1091–1097, https://doi.org/10.1016/j.respol.2019.01.006.

60. Kenneth A. Bamberger and Deirdre K. Mulligan, *Privacy on the Ground: Driving Corporate Behavior in the United States and Europe* (Cambridge, MA: MIT Press, 2015).

61. Angela Huyue Zhang, "The Promise and Perils of China's Regulation of Artificial Intelligence," University of Hong Kong Faculty of Law Research Paper No. 2024/02 (February 12, 2024), http://dx.doi.org/10.2139/ssrn.4708676.

62. One such approach is tentative governance, see Stefan Kuhlmann, Peter Stegmaier, and Kornelia Konrad, "The Tentative Governance of Emerging Science and Technology—A Conceptual Introduction," *Research Policy* 48 (2019), 1091–1097, https://doi.org/10.1016/j.respol .2019.01.006.

63. See, in particular, Anna Christensen, "Normative Patterns and the Normative Field: A Post-Liberal View on Law," in Thomas Wilhelmsson and Samuel Hurri (eds.), *From Dissonance to Sense: Welfare State Expectations, Privatisation and Private Law* (Farnham: Ashgate 1999).

64. Bill C-27, Artificial Intelligence and Data Act, 1st sess., 44th Parliament, 70–71 Elizabeth II, 2021–2022.

65. Personal Data Protection Commission Singapore, *Model Artificial Intelligence Governance Framework*, 2nd ed. (2020), https://www.pdpc.gov.sg/-/media/Files/PDPC/PDF-Files /Resource-for-Organisation/AI/SGModelAIGovFramework2.pdf.

66. "A Pro-Innovation Approach to AI Regulation," UK Department for Science, Innovation and Technology, August 3, 2023, https://www.gov.uk/government/publications/ai-regulation-a-pro-innovation-approach/white-paper.

67. European Commission, *Hiroshima Process International Code of Conduct for Organizations Developing Advanced AI Systems* (October 13, 2023), https://digital-strategy .ec.europa.eu/en/library/hiroshima-process-international-code-conduct-advanced-ai -systems.

68. See, for example, Alex Engler, "The EU and U.S. Diverge on AI Regulation: A Transatlantic Comparison and Steps to Alignment," Brookings Research, April 25, 2023, https:// www.brookings.edu/articles/the-eu-and-us-diverge-on-ai-regulation-a-transatlantic -comparison-and-steps-to-alignment/.

69. INFOCOMM Media Development Authority, "First of Its Kind Generative AI Evaluation Sandbox for Trusted AI by AI Verify Foundation and IMDA," imda.gov, October 31, 2023, https://www.imda.gov.sg/resources/press-releases-factsheets-and-speeches/press-releases /2023/generative-ai-evaluation-sandbox.

70. For a detailed analysis and a discussion of the factors that determine competition or cooperation in standardization efforts, see Nora von Ingersleben-Seip, "Competition and Cooperation in Artificial Intelligence Standard Setting: Explaining Emergent Patterns," *Review of Policy Research* 40, no. 5 (September 2023): 781–810.

71. See John Palfrey and Urs Gasser, *Interop: The Promise and Perils of Highly Interconnected Systems* (New York: Basic Books, 2012).

72. See chapter 10 of Palfrey and Gasser, *Interop*, and expanding on it, Rolf H. Weber, *Legal Interoperability as a Tool for Combatting Fragmentation*, Centre for International Governance Innovation (December 2014), https://www.cigionline.org/static/documents/gcig _paper_no4.pdf.

73. See Weber, "Legal Interoperability as a Tool for Combatting Fragmentation," 9.

74. European Commission, *Hiroshima Process International Code of Conduct.*

75. United Nations General Assembly, *Seizing the Opportunities of Safe, Secure and Trustworthy Artificial Intelligence Systems for Sustainable Development.*

76. See also United Nations AI Advisory Body, "Governing AI for Humanity," Interim Report (December 2023), suggesting ILO's tripartite structure and the UN Global Compact as possible sources of inspiration (p. 16).

77. For early country examples, see, for example "Multistakeholder AI Development: 10 Building Blocks for Inclusive Policy Design," UNESCO and i4Policy (2022).

78. The Global Partnership on Artificial Intelligence, "Multistakeholder Expert Group Annual Report," 2023, https://gpai.ai/projects/.

79. Partnership on AI, "Responsible Practices for Synthetic Media: A Framework for Collective Action," n.d., https://syntheticmedia.partnershiponai.org/.

80. AI Governance Alliance, *Presidio AI Framework: Towards Safe Generative AI Models*, World Economic Forum (2024), https://www3.weforum.org/docs/WEF_AI_Governance _Alliance_Briefing_Paper_Series_2024.pdf.

81. ETH AI Center, "Joining Forces to Reveal and Address the Risks of Generative AI," January 2024, https://ai.ethz.ch/news-and-events/ai-center-news/2024/01/launch-of-a-risk -exploration-and-mitigation-network-for-generative-ai.html#:~:text=As%20a%20 fully%20transparent%20system,developing%20effective%2C%20standardized%20AI%20 testing.

82. In the context of platform regulation, see Chris Riley and Susan Ness, "Modularity for International Internet Governance," *Lawfare*, July 19, 2022.

83. Urs Gasser and Viktor Mayer-Schönberger, *Guardrails: Guiding Human Decisions in the Age of AI* (Princeton, NJ: Princeton University Press, 2024).

CHAPTER 6

Challenges to Evaluating Emerging Technologies and the Need for a Justice-Led Approach to Shaping Innovation

Alex John London

> As it is useful that while mankind are imperfect there
> should be different opinions, so it is that there should
> be different experiments of living; that free scope
> should be given to the varieties of character, short of
> injury to others; and that the worth of different modes
> of life should be proved practically, when any one
> thinks fit to try them.
>
> —Mill, *On Liberty*, Chapter 3

Introduction

Innovation is inherently disruptive.[1] It involves developing or discovering new ideas, new practices, new products, or new services (call these "ends" for convenience), or finding new ways to achieve established ends. It is also deeply social. In some cases, new ends compete with established ends for people's attention or allegiance. In other cases, new ways of achieving the same end compete with old ways of achieving that end. In both cases, the disruptions of innovation can have a profound impact on the rights and well-being of individuals. In some cases, this impact can be positive, as when individuals are better able to more safely or effectively advance ends that are important to them. But the disruptions of innovation can also have negative effects. Not all innovations are successful; some efforts to achieve new ends or to achieve established ends in better ways fail. When unsafe or ineffective technologies circulate, their use can produce direct harms, as when unsafe or ineffective medications subject users to toxic side effects, as well as opportunity costs from not having accessed a safer or more effective alternative. In other cases, successful innovation means that old ends or old ways of achieving established ends are placed at a competitive disadvantage and the people who identify with them, or who built their expertise or life around them, find themselves out of work or displaced in some other way.

A variety of parties whose interests are affected by innovation—individuals, organizations, social institutions, government bodies, policy-makers, lawmakers, and leaders in all sectors—would benefit from an ethical framework that would facilitate the assessment of innovative technologies and of the ecosystem of innovation from which they are produced. Such a framework would be valuable for a variety of reasons, but I focus on three in particular. The first involves *normative guidance*: It would be valuable to the extent that it helps these stakeholders determine when some technological innovation is disruptive but morally permissible, when such disruptions call for some type of social action, and what form such a response should take. The second involves the *allocation of responsibility*: It would be valuable if it could facilitate the process of identifying which agents or actors are responsible for intervening to eliminate, reduce, or mitigate the ethical concerns associated with a particular innovation. The third involves *tracking the health of the innovation ecosystem*: It would be valuable for such a framework to facilitate the assessment of the ecosystem of innovation, understood as the division of social labor and the rules, regulations, laws, social structures, and institutions that shape the process of innovation in order to determine when this ecosystem is functioning in a way that is morally and socially justifiable and when it requires redress or improvement.

In the following section, " Distinctive Challenges to the Ethical Assessment of Innovation," I outline some of the factors that pose a challenge to any such framework. These factors include complexities around understanding or modeling the process of innovation, predicting the effects of innovation, and what the philosopher John Rawls referred to as the fact of reasonable pluralism—the idea that freedom promotes reasonable diversity in moral values and commitments. In " Pragmatic Approaches and the Neglect of Justice," I discuss a common approach to navigating the fact of reasonable pluralism—namely, relying on a "thin" set of ethical principles that might be used to evaluate individual innovations and the innovation ecosystem. These values include the avoidance of harm or nonmaleficence, the provision of benefit or beneficence, and respect for autonomy, fairness, and justice. In " Toward a Justice-Led Approach to Shaping and Evaluating Innovation," I argue that these values are often interpreted in a way that places the greatest emphasis on a set of direct or immediate effects of innovation and that marks out the contributions of a limited set of stakeholders. What is left out is a clear recognition of indirect or higher-order effects from innovation, stakeholders who influence these effects, and the way that these effects can influence considerations of justice. In "Conclusion," I argue that these shortcomings might be mitigated by a framework that adopts a justice-led approach to assessing innovation and the innovation ecosystem.

Distinctive Challenges to the Ethical Assessment of Innovation

The ethical assessment of innovation is complicated by at least three distinctive factors. The first has to do with *freedom and decentralization*. At the most general level, the decision to employ one's intellect, time, and resources in the service of discovering new ends or new means to achieve established ends is morally permissible, if not morally meritorious. It is morally permissible because it falls under the broad liberty to pursue a life plan of one's own. This liberty is itself grounded in two very basic values.

The first is respect for individual autonomy, that is, the ability of individuals to decide how they want to live and to make momentous decisions for themselves is valuable because these freedoms allow individuals to express their individuality, they are central to a person's status as an agent, and they capture a person's interest in exerting fundamental influence, if not control, over how their life goes. Second, the ability of persons to pursue a life plan of their own is fundamental to their well-being—to their ability to lead a life that advances their interests and in which they find satisfaction and fulfillment. Individuals who engage in the process of inquiry, experimentation, and discovery necessary for innovation often do so because such activities are personally rewarding and part of what they regard as a good life.

Beyond being merely permissible, the decision to employ one's intellect, time, and resources in the service of discovering new ends or new means to achieve established ends is often morally meritorious. The reason is that innovation is rarely a purely personal act. As the philosopher John Stuart Mill noted, the knowledge of how to achieve new ends, or how to more effectively or efficiently achieve established ends, often propagates through society so that the benefits produced through innovation are enjoyed by many people. As a result, the process of innovation is often a socially valuable activity to be encouraged.

Academic freedom can thus be seen as a value that sits at the confluence of these two contributories: It protects the rights of individuals to pursue their interests and reflects the idea that, in the aggregate, the free pursuit of novel ideas is likely to contribute to social progress.[2] A legitimate social role for government in an open society, a society in which individuals generally have the liberty to decide how they want to live and how they want to employ their time and energies, is to find ways to manage the risks, costs, and burdens of innovation so that they are fairly distributed and outweighed by the resulting social benefits.

In an open society, the process of innovation is often decentralized. This is not to say that in an open society there will be no efforts to centralize innovation—to facilitate state-sponsored initiatives in science or health—since open societies often do undertake such efforts. It is simply to say that government action will not be the only avenue for innovation and that even when governments are the sponsor of innovation, the process of innovation will often be carried out by entities outside of government. Individuals and associations such as corporations, philanthropies, nonprofits, and other entities can be sources for the discovery of new ends, better means, or for the innovative use of new technologies. Additionally, the process of innovation is not limited to the developers of new technology. Developers may produce a technology with a particular set of goals or uses in mind, but other individuals may use that technology as an occasion for further innovation. For example, the smartphone created a platform for a multitude of developers to create mobile applications, and end users are free to put these devices and their associated software to use in practices that might not have been foreseeable prior to the invention of this platform. If all else is equal, the freedom to experiment and to innovate this way is grounded in the same respect for individual freedom, autonomy, and well-being just discussed.

As a result, the parties involved in innovation can be quite diverse, ranging from individuals, small groups or clubs, to philanthropies, nonprofit organizations, private and public corporations, educational institutions, or entities within local, state, or national government. Some of these parties make decisions as individuals while others make decisions through a complex division of social labor, as when corporations or government bodies make decisions. Some of these parties are also deeply enmeshed in social roles, social structures, or a division of social labor that entails different sets of prior obligations or commitments that guide or constrain their behavior. Likewise, their activities fall into different sectors of social life, from private hobbies to consumer products, individual or public health, employment, banking and finance, criminal justice, security and defense, political participation, the provision of essential social services, and so on. Activities in these different spheres may differ in the ethical issues they raise since they affect different rights and interests of persons or implicate the functioning of social structures with different social functions and expectations.

A second factor complicating the ethical assessment of innovation stems from the degree of *uncertainty* surrounding this process and the difficulty of predicting how it will unfold and what its outcomes will be. Individuals or groups who set out to create or to discover something new often fail and it can be difficult to predict which of their efforts will succeed. Similarly, some efforts at innovation succeed, but not in ways that were originally intended.[3] As a result, innovation is often fortuitous, with efforts to develop something in one area or domain or for one purpose resulting in the ability to achieve some different purpose in a different area or domain. Likewise, it can be difficult to envision how technologies developed to advance one set of goals or purposes might be used in unexpected or innovative ways.

The impacts of innovation are not simply a function of the relationship between a technology and an end user. The emergence of a new technology can alter the way that individuals or groups divide social labor, can shift the nature and function of social roles, and can lead to unforeseen uses that have further impacts on social relationships, opportunity, and the relative costs or ease of performing certain tasks, the relative value of those tasks in a reconfigured environment, and so on. Similarly, the sectors of social life are not static. Innovations in one sector can affect opportunity in others or shift the boundaries between sectors.[4] This in turn can blur lines regarding which set of established norms should be used to evaluate, govern, or regulate a new technology and challenge the utility of the way those norms have been articulated and enforced.

As a result, the interests that are potentially affected by innovation can be extremely diverse. They can include interests that are very specific to an individual because they are tightly bound up with an idiosyncratic feature of their particular life plan, to interests that are widely shared because they are grounded in a human right. Uncertainty surrounding the process and outcomes of innovation entail that these *impacts* can also be difficult to foresee.

A third factor complicating the ethical assessment of innovation stems from *complexity of the relevant normative considerations*. On a very broad level, open societies are characterized by what the philosopher John Rawls refers to as the fact of reasonable pluralism.[5] The basic idea here is that individuals pursue a variety of life plans, often built around a diverse set of "thick" or "substantive" conceptions of the good life. By a conception of the good life, we simply mean a set of goals, values, and ideals that mark out some activities as valuable, worthwhile, or beneficial and others as harmful, ignoble, or lacking in worth.[6] As an extremely simplified example, some people are deeply religious and will forsake wealth or popularity in service to their particular faith tradition; others may regard religion as silly superstition. Some people value music and spend long hours practicing an instrument, whereas others value the exploration of wide-open spaces and would find being cooped up in a room doing the same thing over and over the worst possible existence.

Because different individuals care about, and are committed to, different goals, activities, and ideals, their interests will be advanced or set back by different activities and outcomes. This is centrally relevant to innovation since some people may be deeply invested in, committed to, or may identify with activities or technologies that are displaced by innovation. Individuals who identify deeply with their role in the ice industry, the telegraph, the whale-oil industry, steam engines and the like will find these important interests set back by the development and diffusion of refrigeration, telephony, electricity, and internal combustion or electric engines. For others, the development of these new technologies may be an unalloyed benefit as it enables them to advance more of their interests more effectively and efficiently.

The fact of reasonable pluralism adds to complexities already mentioned surrounding the diversity of the parties involved in innovation, the sectors in which innovation can take place, and the extent to which disruption in these sectors affects interests that are peculiar to individuals or widely shared because they are grounded in some kind of basic human, social, ethical, or legal right. For example, in many countries, health care contexts are governed by a different (usually stricter) set of norms, rules, regulations, or laws compared to consumer products or other business contexts. Likewise, innovations that take place within, or have a significant impact on, relationships between doctors and patients, or lawyers and their clients, individuals and the police, may have different implications from innovations that involve producers and consumers of consumer products.

An acceptable framework for assessing the ethics of technological innovations and the innovation ecosystem should be broad enough to recognize the full range of stakeholders whose activities may be relevant to ethical appraisal, capable of recognizing how social structures mediate social interactions and alter the division of labor and responsibility, and of differentiating disruptions from innovation that are morally permissible from those that rise to the level of an injustice and therefore call for social solutions.

Pragmatic Approaches and the Neglect of Justice

Efforts to develop ethical and policy frameworks to evaluate the process of innovation or the impact of novel technologies have been sensitive to the prospect that they must be capable of providing guidance to stakeholders in a diverse society in which there may be reasonable disagreement over a wide range of issues. This has motivated approaches that are pragmatic in the sense that they do not claim to be grounded in a single, "thick" or substantive, "comprehensive theory" of the good, the good life, the good society, or other set of ethical, social, or political ideals. Instead, proponents appeal to constructs that are supposed to be "thin" or "freestanding," in the sense that they are supposed to have normative force without being tied to and dependent on any single comprehensive theory.

As examples, some have appealed to what they call "common morality," understood as something like a set of pre-theoretical intuitions or commitments that are widely shared and regarded as so important that they need to be accommodated within (rather than overridden or eliminated by) thicker or more substantive conceptions of the good or the good life.[7] A similar idea is that there are certain values that function as "midlevel principles," in the sense that they group and explain a wide range of judgments about particular cases while being common elements within different substantive comprehensive theories.[8] A related concept appeals to what Rawls calls an "overlapping consensus" of reasonable views.[9] Here the idea is that there may be a multiple competing comprehensive theories and that these theories may differ in the way that they justify various claims, but that they often overlap in their endorsement of particular norms or values and the judgments that flow from them.

Although different approaches frame the elements of these thin frameworks slightly differently, they commonly include the following.[10] Nonmaleficence is generally understood as the duty to avoid inflicting harm or imposing burdens on others. Beneficence is the duty to aid, assist, improve, or otherwise benefit others where possible. Respect for autonomy is the duty to respect the interest that other persons who have the capacity to make their own decisions have in being able to make those decisions for themselves. Fairness is the duty to treat like cases alike, to apply the same rules or to follow the same process for all individuals, regardless of features or characteristics that are not directly related to some morally relevant aspect of the case, such as culpability, responsibility, merit, or desert. Finally, justice is widely recognized as an important element in many pragmatic approaches, but its content is often not clear.[11] It is often regarded as a form of fairness, since it involves treating like cases alike and applying uniform procedures or rules, without a clear specification of the grounds to differentiating these two concepts. In many contexts of professional ethics, appeals to these thin or freestanding constructs are bolstered by appeals to role-related obligations of professionals. One of the oldest and most well-developed examples is medical ethics, where the asymmetry of knowledge between doctors and patients, the dependency of patients on doctors, and the profound importance of health to human agency and well-being is seen as grounding a special obligation on the part of doctors to avoid harming patients, to do their best to advance patient interests, and to place those interests above potentially competing interests.

Research ethics is the field that developed to regulate and evaluate the development of new drugs, devices, practices or procedures in medicine. In research ethics, the principles of nonmaleficence, beneficence, respect for autonomy, and justice are codified in *The Belmont Report: Ethical Principles and Guidelines for the Protection of Human Subjects of Research*, a report of the National Commission for the Protection of Human Subjects of Biomedical and Behavioral Research (1979).[12] There are conflicting views about whether the same norms from clinical medicine should also regulate the activities of researchers.[13] Nevertheless, research ethics stands out as a branch of practical ethics that is tightly connected to

a clear set of regulatory requirements and a set of institutions and structures necessary to implement and to some degree even to enforce those
requirements.

Although this pragmatic approach has many virtues, the neglect of justice produces significant shortcomings rooted in the extent to which the
resulting frameworks are highly parochial.[14] In particular, these approaches
mark out as salient an incomplete set of actors, an incomplete set of impacts,
and draw on local norms, often grounded in role-related obligations, to resolve conflicts among its values or principles. For present purposes, the
main point is not to evaluate the merits of the assessments that these frameworks facilitate but to emphasize the types of consideration that such approaches struggle to formulate and to address.[15] To make these concerns
concrete, I illustrate how they apply to research ethics and then consider
how they generalize to the context of machine learning (ML) and artificial
intelligence (AI).

Within research ethics, the dominant focus is on the relationship
between two central parties: researchers and study participants. At the center of this focus is the review of individual study protocols by independent,
local review committees, referred to in the United States as Institutional
Review Boards (IRBs). The purpose of the IRB is to review individual
study protocols, where study protocols basically define the terms on
which researchers will interact with study participants. These interactions
are then assessed according to the set of values described earlier. That is, to
address beneficence, researchers are expected to explain the goals of the
study, the methods that will be used to achieve those goals, the value of the
information that is expected to result from the study (as a proxy for benefits to society), and any benefits that accrue directly to participants from
participation. To address nonmaleficence, they must detail the risks to
which participants will be exposed, the steps that will be taken to eliminate unnecessary risks, to mitigate any remaining risks, and to show how
risks that cannot be eliminated are justified in light of the benefits expected
from the research. To address respect for persons, the protocol must also
contain an account of the information that will be provided to potential
study participants so that they can make a free and informed decision
about whether to participate or not to participate. In cases where this kind

of informed consent is not possible, the protocol must contain a justification for a waiver of consent and specify the steps that will be taken to secure informed consent from a proxy (in cases where participants themselves lack decisional capacity) or to inform participants that they have been involved in research after the fact (e.g., in cases of research on interventions that are used in emergency circumstances). Finally, to address issues of fairness or justice, the protocol must contain a description of the process that will be used to recruit study participants and why this process is fair in the sense of not overburdening populations that are convenient, vulnerable, or easy to manipulate, while taking steps to include populations that are often underrepresented in research.

The system of requiring IRB approval of research before it can be conducted plays an important role in ensuring that abuses of the past are not repeated and helps to provide confidence on the part of study participants that by volunteering to participate in research, they are not submitting to treatment that is unnecessarily risky, abusive, or substantially different from what is described to them during the process of informed consent.[16] Nevertheless, this way of framing the oversight of innovation in biomedicine focuses primarily on direct or *first-order* effects of the interactions of researchers and study participants. Consider now the broad range of issues that are not marked out as salient by this approach.

First, which research questions are asked and how research funds are allocated has a profound impact on which health needs are or are not the subject of investigation. This in turn has a direct impact on whether or not health systems can respond effectively, efficiently, or equitably to the diverse range of important medical needs that are represented in the populations they serve. The current capacity of health systems is the result of long histories of social inclusion and exclusion including histories of oppression and racism but also histories of neglect and indifference. It is also the result of decisions about which health needs to regard as priorities, how to divide social labor for addressing these needs between public health, prevention, and medical care, how research fund should be allocated and what the requirements are for bringing innovative products to market. Call this the *problem of aligning the focus of innovation with the capabilities of social institutions.*

Second, many of the decisions described in the previous paragraph are often not made by researchers but by governmental agencies, such as the National Institutes of Health (NIH) or the National Science Foundation (NSF), nongovernmental funding agencies, philanthropies, or private ventures, such as biotech startups or pharmaceutical companies. Politicians, government employees, and corporate executives are rarely the focus of ethical discussion in research ethics. Yet their decisions have a profound impact on whether a set of basic social institutions—systems that are responsible for individual and public health—have the knowledge and the means to respond safely, effectively, and equitably to the needs of the populations that depend on them and for whether communities perpetuate or rectify health disparities that arise, at least in part from histories of exclusion, animus, or neglect, or abuse. Call this the *problem of full coverage for accountability.*

Third, IRBs evaluating individual trials on a case-by-case basis might regard each study as morally permissible while the portfolio comprised of those studies is morally problematic.[17] For example, the resulting portfolio might be biased to favor the health needs of already advantaged groups, to favor health needs that are traditionally well-studied over health needs that have been neglected, or to advance the pecuniary interests of sponsors without addressing priority health needs of the community. The portfolio as a whole might also expose more participants to worse risks than alternative ways of generating the same information through the application of different study designs. Similarly, the evidence gaps in a portfolio may shift risks and burdens to parties who are already burdened with excessive costs. This problem is partly a consequence of the first two points—the framework in question focuses on an overly narrow set of issues and actors. But it is also a function of the case-by-case approach to evaluation and the absence of guidance for evaluating larger sets of studies and larger strategies of decision-making and the patterns of outcomes or impacts they will produce over time, including the bandwidth of information that can be achieved by different ways of organizing a study portfolio and the evidence gaps that remain. Call this the problem of *portfolio-level ethical issues.*

Finally, each of the preceding points reflects a particular aspect of a more general fact—namely, that innovation takes place within a much larger social ecosystem, one aspect of which is a division of labor among a multiple parties. One function of this division of labor is to shift or transfer the distribution of rights or responsibilities so that there is not a one-to-one correspondence between the actions of a party, the moral appraisal of the outcome that results from that action, and the responsibility to address that outcome. Researchers design and propose individual protocols. But which protocols are funded is a function of the decisions of funding agencies, which, in turn, is influenced by decisions of their leadership, donors, or politicians. A researcher who proposes a study to evaluate a drug in an adult population is performing an act that is morally permissible, if not morally meritorious. Whether that same intervention is ever studied in children is a function of decisions of a much larger set of stakeholders. But the knowledge gap created by a system that does not promote studies in children, pregnant women, or similar populations, can create or perpetuate health disparities with detrimental consequences for the health and well-being of members of these groups. In such cases, although researchers are responsible for the protocols they carry out, responsibility for the ecosystem that shapes the protocols that researchers propose often falls to other parties (e.g., policymakers, funding agencies, drug companies).

More broadly, the advent of new technologies and shifts in their use can cause workers who produced, maintained, or used supplanted technologies to lose their jobs. Developing new technologies is a morally permissible undertaking, as are the general steps it takes to offer a product in a competitive marketplace. But losing a job is a serious setback to a person's interests, constituting harm. Nevertheless, it would be unreasonable to regard this consequence as grounds for holding that the development of innovative technology is morally wrong, and developers of new technologies are not commonly held responsible for these harms or for redress to the workers displaced by it. Rather, responsibility for mitigating the negative consequences of innovation on employment and for facilitating the ability of workers to transition between employment without serious adverse harm usually falls to governments. Call this the *problem of distributed responsibility.*

Interestingly, discussions surrounding ethical and responsible development and use of AI have been sensitive to the fact that these systems can be developed or deployed in ways that recapitulate prior unfairness or injustice. Primarily, this awareness arises because AI systems are trained on large datasets, and these datasets capture patterns in the underlying data-generating process. In a society with sexiest discourse, corpora of text will contain sexist language. In a society with racist histories, the groups marginalized by such attitudes and practices will be underrepresented in databases generated from the provision of medical care or other social services and overrepresented in databases used to police or penalize. Likewise, databases will contain demeaning or racist statements about groups that are subject to social animus and reflect associations between certain traits or characteristics and attitudes of normality versus aberrance, beauty versus ugliness, and competence versus incompetence. Training AI systems on this data can perpetuate these judgments and attitudes.[18] Recognizing these relationships and taking steps to effectively manage, mitigate, or eliminate these biases is extremely important.

Because these biases can be inherited from training data, the responsibility for managing them is often seen as falling on the shoulders of developers. The problem is also framed in relatively narrow terms of discordance between training data and the ground truth in the relevant population. As a result of these assumptions, the vast majority of the burgeoning literature on fairness in AI focuses on statistical properties of model outputs, such as the relationship between false negatives and false positives along with a guiding assumption that the relevant considerations of fairness or justice are local—they have to do with the rules that should govern the distribution of specific goods, opportunities, or services.[19] Against this background, the central assumption is that developers should ensure that each person receives equal treatment relative to this set of standards for local justice.

One problem with this focus on local justice derives from what I called the problem of portfolio-level ethical issues: Each algorithm evaluated on a case-by-case basis, evaluated solely for their conformity to considerations of local justice, might be morally acceptable, but the system of such algorithms could be deeply unjust. This is possible because society is not just a collection of interactions that operate independently of one another. It is, rather, a network of interrelated interactions, often mediated by social institutions that affect overlapping aspects of people's opportunities, capabilities, rights, and interests. As a result, historical injustice in one domain, such as housing,[20] finance,[21] or policing,[22] can have a profound, detrimental impact on the health of oppressed populations, the quality of education available to them, their ability to take advantage of educational opportunities, their career prospects, their ability to vote or hold political office, their freedom to move and associate, their financial prospects, and other important rights and interests. Prior injustice in one aspect of society creates disparities that reduce or impede the opportunities or capabilities of affected individuals or groups. When this is the case, norms of local justice in other parts of society can effectively ensure that disadvantaged populations remain at a disadvantage in transactions or relationships that take place in those domains.

As a result, upholding norms of local justice in the operation of important social institutions (such as access to education, opportunities for employment, and so forth) can serve to reinforce unjust disparities and social inequalities that arise from prior histories of unfair treatment. The myopic focus of local justice is poorly suited to the task of recognizing injustice in the operation of larger social structures (to the problem of structural injustice) and to framing strategies for enacting justice as rectification—the process of rectifying unjust practices and mitigating their effects on disadvantaged parties with the goal of restoring relationships of equal standing, equal regard, and fair treatment.

The dynamics outlined in this section illustrate important shortcomings in frameworks that focus primarily on developers or firms that develop particular technologies, the impact of particular technologies on users or the targets of the technology, and on issues of fairness that are framed as complying with the norms for local justice.

Toward a Justice-Led Approach to Shaping and Evaluating Innovation

The neglect of justice in practical ethics stems, at least in part, from the perception that every formulation of this value is necessarily tied to and embodies some thick, comprehensive conception of the good, the good life, or the good community and that, therefore, it is incapable of securing the kind of widespread commitment necessary to guide policy in an open, pluralistic society. This concern is not without merit since there certainly are competing and potentially conflicting comprehensive conceptions of justice to which some people are deeply committed. But this prospect should not be a deterrent to identifying *elements of justice* that can make salient the ways in which innovation and innovations can affect important social institutions, relationships, opportunities, or interests. Making these issues salient means not only drawing attention to them but highlighting reasons why they may need to be addressed and helping to identify which stakeholders might have responsibility for redress. Such a framework need not provide complete solutions to the problems it identifies. But we cannot solve problems we do not formulate, and being able to formulate the ways in which innovation and innovations might raise concerns of justice can facilitate concrete action, even if this must play out within some larger political process.

A justice-led approach would begin by identifying the space within which diverse members of an open society have a claim to equal standing and equal regard. The idea that justice is fundamentally concerned with giving equal treatment to equals, and treating like cases alike, requires a specification of the respect in which individuals are equal and in which they have a claim to like treatment. The fact of diversity entails that individuals in an open society embrace and follow different substantive, first-order conceptions of the good. But amid this diversity, every such individual should also recognize that they share a higher-order interest in having the real freedom to formulate, pursue, and revise a life plan based on some first-order conception of the good. This shared higher-order interest need not be grounded in or tied to any particular conception of the good. It can be grounded solely in the recognition that there is a more general respect in which each person in a diverse, open society is engaged in the same kind of fundamental project (formulating, pursuing, and revising a life plan that embodies some set of ideals and values) and that this project is of deep personal and social importance to each of those individuals.

This shared interest constitutes a compelling ground for claims of equal standing and equal regard. First, it captures a social perspective that is available to, and that has a compelling rational claim on, every individual in a diverse and open society. Different individuals embrace different values, goals, and ideals, but they can see one another as engaged, at a more general level, in a shared project that is of profound importance to each individual. Second, from this higher-order perspective there are no grounds on which to regard any individual, or set of individuals, as in any respect better than, superior to, or more deserving than any other. Individuals who pursue different life plans, and so hold different values, are nevertheless equal in the sense that they each want to be free to formulate, pursue, and revise some first-order life plan. Third, this social perspective is consistent with and can accommodate all first-order life plans that are reasonable in the following minimal respect: They are not predicated on the domination or subordination of some other group or class of persons. This notion of reasonableness is not grounded in a thick conception of reason but follows simply from the recognition that societies are constituted by distinct individuals, that every individual shares an interest in having real freedom to formulate, pursue, and revise a life plan of their own, and that at this more general level there are no grounds for regarding one individual as superior to or as having any right to dominion or priority over any other.

Next, a critical role for, and criteria for the justification of, social institutions in a diverse, open society is to create and maintain conditions that secure and promote this shared interest. This includes institutions of government and security that affect the distribution of rights, privileges, and prerogatives as well as institutions that influence the distribution of social opportunity and material resources such as the institutions of individual and public health, provisions for a social safety net, and institutions that govern employment and market-based transactions and relationships. Institutions that secure and promote this shared interest can be seen as supporting the ability of these distinct individuals to function as free and equal persons.[23]

This focus provides *normative guidance* that can help stakeholders determine when the positive or negative impacts of the ecosystem of innovation or of specific innovative technologies is disruptive but not unjust and when these disruptions rise to the level of an injustice. First, because a key function of basic social institutions is to uphold conditions that respect the moral equality of persons by securing their shared higher-order interest in having the real freedom to formulate, pursue, and revise a reasonable life plan, these social institutions should be called into action when individuals face widespread threats to this shared higher-order interest. The diffusion of innovative technologies can alter social conditions that affect this higher-order interest. In such cases, just social institutions should intervene to promote equal treatment and equal regard.

As an example, successful innovation often creates social circumstances in which some individuals can advance their personal ends more effectively or efficiently than others and this necessarily creates inequalities. When these advantages or disadvantages are limited to advancing or detracting from a person's specific first-order life plan, then those advantages or disadvantages qualify as benefits or harms and fall under the rubric of.[24] If this is all that is at stake, then these benefits or harms do not rise to the level of an injustice. The reason is that justice is not concerned with how well individuals are able to achieve the specific first-order life plan they set out for themselves—this is the domain of beneficence. Rather, justice concerns the higher-order interest of individuals in having real freedom to formulate, pursue, and revise some reasonable life plan.

Treating such inequalities as unjust per se would require that we refrain from producing innovative technologies unless they can advance the first-order life plans of all individuals equally. But this is likely an impossible requirement, since the diversity of life plans frequently involves zero-sum relationships involving rival goods—goods that cannot be enjoyed by multiple agents simultaneously. These include positional goods (e.g., being the best at something) and other scarce resources. Promoting equality by preventing advances unless those advances benefit everyone to the same degree relative to their distinctive life plan would require that we secure equality by "leveling down," which is to say, it would make some people worse off, without the prospect of making anyone better off, simply to ensure their equality to others. In other cases, innovation creates inequalities that directly influence the higher-order interests of individuals because the knowledge or the means that are produced cannot benefit all persons equally. For example, advances in cancer research might extend the lives of or restore physical functioning to patients with one type of tumor but not to all cancer patients. Similarly, there are cases where researchers first seek to establish that some technology works in what is regarded to be a comparatively easy test case before trying to extend its use to a wider range of applications. For example, hemophilia might be an excellent model system for gene-based therapeutics if it is believed to represent a comparatively simple application of a new technology. Initial successes in treating this condition would generate inequalities, since gene-based treatments for other conditions would not be available in the same time frame.

Here again, promoting equality by prohibiting research that would save some lives just because we would not know how to save all others would be self-defeating. If all else is equal, stepwise innovation in which developers seek to unlock the benefits of an intervention in one case and then to extend it to others is morally permissible. The problem, however, is that frequently, all else is not equal.

In particular, this is another *portfolio-level ethical issue.* Problems arise when the portfolio of such decisions winds up tracking, and therefore, recapitulating histories of social exclusion and marginalization—as when research systematically excludes women or members of marginalized groups or when research does not focus on health needs that are distinctive of such groups. Problems also arise when research is undertaken under the assumption that successful efforts will be followed by further research only to see further research not carried out. This happens in drug development, for example, when new interventions are tested in adults first but then subsequent trials in pediatric populations are not carried out.

Similarly, the proliferation of machine learning and artificial intelligence raises concerns about justice because, as we noted earlier, data on which AI systems are built often reflect patterns of social interaction in which specific groups have been subject to unfair treatment. The widespread use of this data creates models that can recapitulate these patterns of social inequality. The point here is that even when individual AI systems are developed to function in ways that primarily are valuable to individuals relative to their individual life plan, concerns of justice arise because social bias is widespread and therefore likely to affect a broad range of datasets used to develop such systems and because the impacts of such problems are connected to histories of exclusion and subordination. The widespread recapitulation or exacerbation of these histories of exclusion and subordination creates an issue of justice because this adversely affects the higher-order interest of affected groups in being treated as free and equal members of society. There is, thus, a strong social interest in eliminating these disparities that applies across the full range of areas where these applications might be applied. It also generalizes beyond ML and AI. Disparities in technologies that adversely affect individuals from groups that have historically been subject to neglect, animus, exclusion, domination, or subordination threaten to recapitulate or exacerbate relationships that are antithetical to justice. Widespread acceptance of these disparities signals that some individuals have lower standing or status than others—a message that is also antithetical to justice. When social institutions act to reduce these disparities, it advances an important cause of justice—ensuring that all people are treated as free and equal.

Second, the critical role of basic social institutions in securing this higher-order interest of individuals combined with the diversity of needs and of circumstances entails a legitimate social interest in promoting innovation that can increase their effectiveness, efficiency, and equity. This provides normative guidance relevant to the *problem of aligning the focus of innovation with the capabilities of social institutions*. The importance of the ability of these institutions to function equitably follows from the fact that, relative to this shared higher-order interest, there are no grounds for regarding the needs of any individual or group as somehow superior to or more important than the needs or interests of any other individual or group. As such, these institutions should strive to function with equal efficacy for all members of the population they serve. However, histories of racism, ableism, sexism, and other forms of social animus, marginalization, or exclusion have created unfair social disparities as well as deficiencies in the functioning of these institutions that exacerbate these disparities. There is thus a strong social and moral imperative to rectify these disparities and to promote the development of technologies that better enable important social institutions to promote the real freedom of all individuals and groups.

The social imperative to ensure that these institutions function effectively and efficiently is grounded in scarcity—the shortfall between resources available and the needs of individuals or social groups—and from the critical role that these institutions play in promoting the real freedom and equality of the individuals whose lives they affect. It follows from these first two points that there is a strong social imperative to ensure that the dissemination and incorporation of new technologies does not undermine or detract from the ability of these social systems to function effectively, efficiently, and equitably.

Here again, ML and AI have been used in applications that have negatively affected the capacity of basic social institutions to function. In this case the resulting disparities affect the functioning of social institutions that have an immediate and direct impact on the ability of individuals to function as free and equal persons. Disparities in algorithms used in policing, sentencing, bail, or parole decisions are unjust because of the strong claim of each individual to equal standing and equal regard in this space. The same is true for disparities from AI systems that make decisions regarding employment, lending, banking, and the provision of social services. Such disparities are unjust even if they are not connected to prior histories of exclusion, indifference, or subjugation because of the important role that the decisions and conduct of these social institutions play in securing this higher-order interests of persons. But such disparities can be, and often are, doubly concerning precisely because they are connected to, and do recapitulate or compound, prior histories of subjugation.

Third, this approach provides a framework for addressing the *problem of full coverage for accountability* because it situates the activity of innovation in a larger web or network of social relationships among a broader set of individuals and groups, highlighting the role of different stakeholders in shaping the process of innovation and allowing for a more explicit consideration of the appropriate division of social and moral responsibility between these parties. This includes the relationship between developers, funders, regulators, policymakers, users, and various social institutions that are required to serve specific social functions grounded in considerations of justice. All individuals in a society rely on these institutions to support and protect their higher-order interest in being able to formulate, pursue, and revise a life plan; these social institutions are sometimes called into action to support innovation (as when government agencies sponsor and support innovation directly, or when they carry out regulatory or legal functions that shape the incentives of actors in this ecosystem); and these institutions are affected by the process and outcomes of innovation, as when their capacity to perform their functions depends on the capabilities of the technologies they deploy for this purpose.

Responsibility for identifying shortfalls in the capacity of important social institutions to secure and promote this higher-order interest for all community members falls to government leaders and to leaders in the relevant social institutions, in consultation with community members. This includes identifying threats to this higher-order interest—from sickness, injury, and disease, environmental degradation, and hazards; to access to employment, social, economic and political opportunity, and social limitations imposed by the built environment; from social animus, exclusion, or indifference; and from the way that novel technologies might unduly consolidate social or political power. These stakeholders also bear primary responsibility for identifying broad priorities for investing in innovation and development with the goal of reducing or eliminating these shortfalls and addressing these threats. This includes identifying and rectifying social inequalities that undermine the freedom or equality of individuals or groups including inequalities that stem from prior histories of animus, indifference, neglect, or other forms of domination or marginalization.

As an illustration of how this framework makes more tractable the *problem of distributed responsibility*, this framework recognizes that individual innovators have broad liberty to pursue the ideas, programs, and projects that interest them. This follows from respect for the freedom of individuals and groups to pursue a reasonable first-order conception of the good and from the difficulty of identifying which avenues of innovation will succeed and how they might be taken up and adopted in innovative ways by others. To align this liberty with considerations of justice, policymakers, regulators, funders, and other leaders have a responsibility to create *incentives* that encourage individuals to explore avenues for innovation that connect to and address knowledge or capability gaps within these priority areas. This responsibility is not widely recognized, and it is a virtue of the current approach that it would make salient the responsibility of this wider range of actors and facilitate collaborative efforts to advance these important social goals.

Similarly, when it comes to identifying and averting portfolio-level ethical issues, the present framework identifies social and political leaders, in conjunction with community members and the heads of entities that fund research or carry out innovation, to identify when patterns of local decision-making can recapitulate, exacerbate, or create patterns of social exclusion or marginalization, and then to intervene, whether through rules or incentives, to rectify such patterning. This can involve ensuring that the novel technologies that address distinctive needs of marginalized or minoritized groups are equitably funded, ensuring that novel technologies are developed in populations that include such groups, ensuring that novel technologies are extended to use cases that affect such groups and ensuring that sequential strategies for testing or development are funded and carried to fruition.

These examples represent problems that range beyond the purview of individual researchers. They arise because of the potential for repeated decisions made solely on myopic criteria to recapitulate or exacerbate larger patterns of inequality, and they call for attention from a wider range of stakeholders including policymakers, research funders, and regulators. It is a virtue of the present framework that it can make salient such second-order issues and facilitate the identification of parties in the innovation ecosystem who should bear responsibility for addressing these issues. This, in turn, can help lawmakers, policymakers, civic and corporate leaders, activists, and other community members craft rules, policies, norms, and incentives that discourage activities that threaten to undermine the equal standing of individuals and promote activities that enhance the ability of social institutions to secure and to promote this shared higher-order interest.

Finally, the approach outlined here provides high-level benchmarks that stakeholders might use to assess the relative health of the innovation ecosystem and the range of norms, rules, practices, regulations, and laws that constitute its governance structure. In particular, this ecosystem is healthier if it has a governance structure that addresses the set of problems outlined here. In other words, innovation ecosystems are healthier to the extent that their governance structure identifies the full range of stakeholders with responsibilities in this area to ensure *full coverage for accountability* for the purpose of protecting the higher-order interests of persons and *aligning the focus of innovation with the capabilities of social institutions*. Similarly, the various incentives that influence the conduct of these agents should ensure that *responsibilities are distributed* to relevant parties and then enforced in a coherent manner, so that *portfolio-level ethical issues* can be identified and addressed.

Conclusion

This chapter outlines a justice-led approach to evaluating the innovation ecosystem and the innovations that it produces. The proposed framework is pragmatic in the sense that it is grounded on moral claims that should have wide purchase on diverse members of an open society without requiring special commitment to some particular conception of the good, the good life, or the good community. It articulates a respect in which members of a diverse, open society can see one another as free and equal, and it recognizes the special role that social institutions play in upholding this conception of freedom and equality. This position is consistent with broad respect for individual and academic freedom while also outlining mechanisms that can be used to ensure that the division of social labor serves to expand the capacity of important social institutions to protect and advance the shared interest of those who depend on them. This breadth of scope creates a framework in which the activities and responsibilities of a broader range of agents can be articulated and evaluated. It is also not limited to the direct or first-order effects of specific agents on others. It can recognize impacts that arise from the cumulative or synergistic interactions of portfolios of decisions.

Clearly, this sketch requires additional work to flesh out key details and improve its relevance to policy. However, the framework outlined here is likely to be particularly sensitive to growing concerns about the impact of AI systems on democratic accountability, public discourse, the integrity of elections, and the role of science and evidence in democratic governance. The reason is that the role of government and the critical social institutions of government, within this framework, is to secure the higher-order interest that all persons share in having the real freedom to formulate, pursue, and revise a life plan of their own. When technologies proliferate in ways that threaten the ability of citizens to hold political leaders accountable, to identify truth from fabrication, to ensure the integrity of elections, and to participate in democratic deliberation, these impacts implicate issues of justice. Moreover, these impacts need not be tied to individual developers and their individual technologies. They can arise from the synergistic interactions of a multitude of novel technologies and from the conduct of stakeholders including corporate executives, lawmakers, politicians, and ordinary people who use and abuse technology. This is an important area in which delineating conduct that is disruptive but morally permissible from conduct that is disruptive and morally problematic is particularly pressing. It is unlikely that such distinctions can be fruitfully drawn and defended without appeal to considerations of justice.

Notes

1. I use the term *disruptive* in a colloquial sense meaning to bring about change. This differs from the more restricted sense in Clayton M. Christensen, *The Innovator's Dilemma: When New Technologies Cause Great Firms to Fail* (Cambridge, MA: Harvard Business Review Press, 2013). On different uses of this term, see Steven Si and Hui Chen, "A Literature Review of Disruptive Innovation: What It Is, How It Works and Where It Goes," *Journal of Engineering and Technology Management* 56 (2020): 101568; and Jeroen Hopster, "What Are Socially Disruptive Technologies?" *Technology in Society* 67 (2021): 101750.

2. Vannevar Bush, *Science, the Endless Frontier* (Princeton, NJ: Princeton University Press, 2021).

3. Royston M. Roberts, *Serendipity: Accidental Discoveries in Science* (Hoboken, NJ: Wiley, 1991).

4. Debra J. H. Mathews, Rachel Fabi, and Anaeze C. Offodile II, "Imagining Governance for Emerging Technologies," *Issues in Science and Technology* 38, no. 3 (2022): 40–46.

5. John Rawls, *Political Liberalism* (New York: Columbia University Press, 1991); John Rawls, *Justice as Fairness: A Restatement* (Cambridge, MA: Harvard University Press, 2001).

6. Alex John London, *For the Common Good: Philosophical Foundations of Research Ethics* (Oxford, UK: Oxford University Press, 2021).

7. Tom L. Beauchamp and James F. Childress, *Principles of Biomedical Ethics*, 5th ed. (Oxford, UK: Oxford University Press, 2001); Bernard Gert, *Common Morality: Deciding What to Do* (Oxford, UK: Oxford University Press, 2004); Bernard Gert, Charles M. Culver, and K. Danner Clouser, *Bioethics: A Return to Fundamentals* (Oxford, UK: Oxford University Press, 2006).

8. Tom L. Beauchamp and James F. Childress, *Principles of Biomedical Ethics*, 8th ed. (Oxford, UK: Oxford University Press, 2019).

9. Rawls, *Political Liberalism*; Rawls, *Justice as Fairness*.

10. Luciano Floridi et al., "AI4People—An Ethical Framework for a Goof AI Society: Opportunities, Risks, Principles, and Recommendations," *Minds and Machines* 28 (November 26, 2018): 689–707; Anna Jobin, Marcello Ienca, and Effy Vayena, "The Global Landscape of AI Ethics Guidelines," *Nature Machine Intelligence* 1 (September 2, 2019): 389–399; Onur Bakiner, "What Do Academics Say About Artificial Intelligence Ethics? An Overview of the Scholarship," *AI and Ethics* (2022): 1–13.

11. Alex John London, "The Independence of Practical Ethics," *Theoretical Medicine and Bioethics* 22 (2001): 87–105.

12. National Commission for the Protection of Human Subjects of Biomedical and Behavioral Research, *The Belmont Report: Ethical Principles and Guidelines for the Protection of Human Subjects of Research*, Department of Health, Education, and Welfare (April 18, 1979), https://www.hhs.gov/ohrp/regulations-and-policy/belmont-report/read-the-belmont-report/index.html.

13. Paul B. Miller and Charles Weijer, "Rehabilitating Equipoise," *Kennedy Institute of Ethics Journal* 13, no. 2 (June 2003): 93–118; Franklin G. Miller and Howard Brody, "A Critique of Clinical Equipoise. Therapeutic Misconception in the Ethics of Clinical Trials," *Hastings Center Report* 33, no. 3 (2003): 19–28; Paul B. Miller and Charles Weijer, "Fiduciary Obligation in Clinical Research," *Journal of Law, Medicine, and Ethics* 34, no. 2 (2006): 424–440; Paul B. Miller and Charles Weijer, "Trust Based Obligations of the State and Physician-Researchers to Patient-Subjects," *Journal of Medical Ethics* 32, no. 9 (2006): 542–547; Franklin G. Miller and Howard Brody, "Clinical Equipoise and the Incoherence of Research Ethics," *Journal of Medicine and Philosophy* 32, no. 2 (2007): 151–165; see also London, *For the Common Good*.

14. London, *For the Common Good*.

15. In parallel work we have explored how the value of beneficence can be understood in a narrow sense that relates to the ability of a technology to expand the ability of users to do something they could not do before, and in a meaningful sense, in which a technology expands the ability of the user to function in ways that they value. A. J. London and H. Heidari, "Beneficent Intelligence: A Capability Approach to Modeling Benefit, Assistance, and Associated Moral Failures Through AI Systems," *Minds and Machines* (forthcoming).

16. A. J. London, "A Non-Paternalistic Model of Research Ethics and Oversight: Assessing the Benefits of Prospective Review," *Journal of Law, Medicine and Ethics* 40, no. 4 (2012): 930–944.

17. Alex John London and Jonathan Kimmelman, "Clinical Trial Portfolios: A Critical Oversight in Human Research Ethics, Drug Regulation, and Policy," *Hastings Center Report* 49, no. 4 (2019): 31–41.

18. Jesutofunmi A. Omiye et al., "Large Language Models Propagate Race-Based Medicine," *NPJ Digital Medicine* 6 (October 20, 2023): 195, https://www.nature.com/articles/s41746-023-00939-z.

19. Jon Elster, *Local Justice: How Institutions Allocate Scare Goods and Necessary Burdens* (New York: Russel Sage Foundation, 1992); H. Peyton Young, *Equity: In Theory and Practice* (Princeton, NJ: Princeton University Press, 1995).

20. Richard Rothstein, *The Color of Law* (New York: Liveright, 2017).

21. Mehrsa Baradaran, *The Color of Money* (Cambridge, MA: Harvard University Press, 2017).

22. Michelle Alexander, *The New Jim Crow* (New York: New Press, 2012).

23. London, *For the Common Good.*

24. London and Heidari, "Beneficent Intelligence."

CHAPTER 7

Bringing Power In: Rethinking Equity Solutions for AI

Shobita Parthasarathy and Jared Katzman

There is great hope that artificial intelligence (AI) and machine learning (ML) can benefit society, from providing real-time translation to more accurate cancer screening. But there are also growing concerns that it is exacerbating social inequity and injustice. In recent years, media reports have revealed the serious negative consequences of the biases in AI datasets, including false arrests triggered by facial recognition technology.[1] Meanwhile, despite the hope that AI will help criminal court judges reduce bias, its use seems to amplify structural inequalities in the justice system.[2] The workers training algorithms to ameliorate bias receive little pay and labor under extremely stressful conditions.[3] At the same time, AI tools meant to benefit marginalized communities are often inaccessible to them.[4]

In response to these emerging equity and justice concerns, policymakers, academics, and the technical community have proposed solutions. The Blueprint for an AI Bill of Rights developed by the Biden administration recommends identifying statistical biases in datasets; designing systems to be more transparent and explainable in their decision-making; incorporating proactive equity assessments into system design, including input from diverse viewpoints and identities; ensuring accessibility for people with disabilities; predeployment and ongoing disparity testing and mitigation; and clear oversight.[5] Scholars have suggested new evaluation capabilities for existing government agencies and even the creation of new regulatory structures.[6] In parallel, the technology industry has focused on educating programmers about the impact of social biases on AI software and creating a market for fairness monitoring tools and services.[7]

These initiatives will surely address some harms. However, most do not address the social inequalities that shape the landscape of technology development, use, and governance, including the concentration of economic and political power in a handful of technology companies and the systematic devaluation of lay contributions and perspectives, especially from those who have been historically marginalized. As a result, the proposed solutions are likely to fall short. To establish a better AI innovation ecosystem and more equitable and just technologies, we must develop solutions that account for these historical inequalities and power imbalances, in addition to addressing current concerns like bias and discrimination in model predictions.

Current Approaches to AI Inequities

Efforts to make AI more equitable rest on the growing realization that many communities, particularly those who have been historically marginalized, have not benefited—and some have been harmed—by technology.[8] By *equity*, we mean "the overarching driver of a process for identifying and ameliorating structural and social conditions that disadvantage individuals and groups by unfairly limiting their freedom, their opportunity, or the conditions needed for well being."[9] In some cases, the problem is simply one of access. Prospective users may not be able to afford a technology or it is otherwise unavailable.[10] In others, the problem is one of design: developers build a technology without understanding the needs and characteristics of a user community, and sometimes even with biased assumptions.[11] This can ultimately have deleterious effects on communities that are already disadvantaged, especially because these assumptions and values are hidden in technical specifications. Finally, there are inequities even early in the process, in terms of who gets to set priorities and how data that informs development is gathered and categorized.

AI equity solutions fit into four categories: technical, organizational, legal/policy, and enhancing civic capacity. *Technical solutions* understand equity in terms of accuracy and focus on reducing disparities in model performance (often referred to as "algorithmic bias").[12] Technologists recognize that when an AI system does not perform equally for all groups of people, it can produce social exclusion, as when a Black smartphone user struggles to unlock their device using facial recognition technology,[13] or when they cannot wash their hands in a public bathroom because the sensor does not recognize their skin.[14] To address such problems, developers try to improve the data, software, and other technical dimensions of a system's design. They may refine datasets to better represent diverse social groups, optimize AI algorithms to mitigate social biases, and implement more stringent quality assurance through additional testing. Inequalities become technical errors. For instance, when Joy Buolamwini and colleagues discovered that major facial recognition platforms had difficulties identifying darker-skinned individuals and women,[15] many companies responded by collecting photos featuring more dark-skinned and female faces and retraining algorithms to take this additional data into account.[16] In some cases, researchers may ask marginalized groups to provide input during the development process but give them no meaningful power to shape priorities or influence results.[17] This alienates these communities further.

Such solutions can only have limited impact because the datasets themselves are assembled in a structurally biased context. Algorithms designed to predict crime, for example, use historical data that reflect discriminatory policing practices.[18] As a result, they tend to overpredict crime in communities of color that are already over-surveilled. And, characterizing data as the solution produces perverse incentives that can exacerbate disproportionate burdens. A contractor working for Google tried to fix inaccuracies in facial recognition technology by paying unhoused Black men in Atlanta a few dollars each to play with a phone.[19] The phone took pictures of the men to improve the datasets without their informed consent. Then even when the AI is technically accurate, it can produce unjust outcomes. Cities have used facial recognition cameras to curb the freedoms of residents in public housing, many of whom are Black, by aggressively surveilling and policing them.[20]

Organizational solutions deployed across the development process view people, practices, and programs as the route to achieving equity. Often used by the tech industry, they include initiatives to make the workforce more diverse and inclusive, "responsible AI" offices identify ethical principles to guide research and development by training technologists about the ethical and social dimensions of their work, supporting humanistic and social scientific research related to AI, and projects that bring user needs explicitly into product design. Microsoft, for example, has issued multiple iterations of its "Responsible AI Standard" to guide technology development across its organization.[21] The Responsible Computing Challenge funded by the Mozilla Foundation trains the next generation of technologists to think holistically about technologies, considering them in social and political contexts.[22]

However, these solutions are often seen as auxiliary to the main project of technology development and are therefore dismissed. Ethics teams inside companies tend to be the first fired when the industry contracts economically.[23] Even when the industry is stable, these teams lack resources, authority, and ultimately impact.[24] Consider the now-famous case of Google firing Timnit Gebru and Margaret Mitchell, who led the company's ethical AI group. Google tried to suppress a paper they coauthored, which discussed environmental harms and racial, gender, and other biases triggered by large language models.[25] When the pair refused, the company fired them.[26] Similarly, despite evidence that the rise of AI will place enormous strain on electricity and water, disproportionately burdening marginalized communities,[27] technologists tend to exclude such factors from their definition of responsible AI and ethical practices.[28] After all, considering environmental impacts would raise the question of whether AI can ever be responsible. Ultimately, because the public is completely dependent on how technologists define responsible, ethical, and equitable AI, they can become little more than buzzwords.

The success of organizational solutions also depends on institutional culture. Even as they have increased their efforts to diversify their ranks, tech companies have struggled to retain employees of color due to alienating work environments. One Black Facebook recruiter has recounted insensitive comments and stereotyping in discussions about hiring,[29] while a Black Google recruiter reported inadequate pay and promotion opportunities for people of color.[30] If the dominant communities in an organization are not reflective or open to change, it is impossible for a handful of employees or new initiatives to produce equitable or just outcomes. Organizational culture also matters in AI use: Although the FBI has a training program for law enforcement officials who use facial recognition technologies, only 5 percent have taken it.[31]

Due to growing concern that tech companies cannot be trusted to police themselves, scholars, civil society groups, and even some technologists have turned to governments for help. *Legal and policy solutions* include temporary moratoria and bans on specific applications, requirements for companies to disclose AI use in "high-risk" decisions, and new government capabilities to assess the effectiveness of AI products. New York City passed a first-of-its-kind law in 2021 to regulate AI use in hiring practices. It requires companies to work with independent auditors to evaluate, on an annual basis, whether their tools exhibit bias in hiring decisions based on race or gender. Job candidates also have the right to request data collected about them. The European Union's pending AI Act will establish a regulatory approval process for technologies it deems high risk, including for migration, asylum, and border control management and biometric identification.

Such laws represent a significant step toward addressing AI inequities by reducing inaccurate uses, preventing disparate impacts, and protecting civil liberties. However, they are often vague and difficult to enforce. The New York law does not adequately define its auditing requirement.[32] As a result, AI companies will have financial incentive to seek lenient bias assessments, and auditors, facing market pressures, will have little leverage to produce more critical and thorough reports.[33]

In addition, governments often justify moratoria and bans on the basis of perceived technical inaccuracies. Cities have banned facial recognition technology because of its poor performance among marginalized communities.[34] But this does not grapple with the civil rights and liberties questions. For example, is it appropriate to allow facial recognition technology in neighborhoods that have suffered for generations due to excessive surveillance?

Last, there are emerging efforts to enhance *civic capabilities*, empowering the public to participate in discussions and even decisions regarding AI. This includes new social movements and new institutions to gauge, engage, and explicitly serve public priorities. The Ford Foundation's Technology and Society Program tries to encourage a vibrant civil society surrounding digital technology, funding the Center for Democracy and Technology, the Leadership Conference for Civil and Human Rights, the People's Tech Project, and the Distributed AI Research Institute. The United Kingdom's Ada Lovelace Institute regularly conducts public dialogues on topics that include the responsible use of location data, trustworthiness of data-driven public health responses, and the use of biometric identification technologies including facial recognition.[35] The US National Science Foundation has tried to create the National Artificial Intelligence Research Resource to broaden access to AI development resources.[36]

This has enhanced public discussion and produced important critiques. But these organizations receive very little funding compared to the investment in technology design,[37] and as a result they are often spread quite thin and end up chasing after individual technologies rather than first imagining the society they want and then considering the role they want technology to play. Many also focus on representing the public as a whole, which means that they may be less adept at identifying issues of specific concern to marginalized communities. Finally, because these efforts are almost always institutionally separate from technology development and policy making, their impact is limited. Consider, for example, the ongoing discussion about the potential for existential risk from AI, initiated by a letter signed by the CEOs of major technology companies in March 2023. This frustrated civic tech leaders, who for years have called attention to the harms and inequities, including algorithmic bias, already produced by AI.[38] But in comparison to the worries about existential risk, their concerns have had little impact. In fact, some civic tech leaders signed the existential risk letter so that they could bring some attention to their concerns, not because they worried that AI would kill us all.[39]

Structural Inequity in Science and Technology

The aforementioned efforts to address equity are serious and well-meaning, but by and large they do not take into account the historical power imbalances that mark the AI ecosystem. As a result, they are likely to have limited impact. In particular, we point to two things: the economic, political, and epistemological influence of technologists and the tech industry; and the systematic discrimination some communities have faced in science and technology for generations. Both, we argue, shape what counts as an equity problem, what counts as a solution, who participates, and how they do so.

Technologists have long had an authoritative role in Western societies. In the early days of the United States, the founders saw the development of new inventions as key to the country's prosperity.[40] This enthusiasm only accelerated in the twentieth century, after the Manhattan Project demonstrated that scientists and engineers could produce technologies for the national interest.[41] Western governments began to increase their investments in research and development and to view innovation as the route not only to national security and economic competitiveness but also social progress.[42] As the technology industry began to grow, then, it was naturally the object of great pride and fascination. Microsoft, Google, Apple, Amazon, and Facebook were not only creating products that the public seemed to want, but they had significant economic value. Excited by their potential, governments were reluctant to hear concerns or regulate them.[43] They have since become so dominant that they are known as "platforms," where they control multiple markets and the behavior of other companies.

This has created concentrated economic power: in 2023, eight of the ten richest people in the world made their money in tech, and the six big tech companies accounted for nearly all of the S&P 500's return.[44] Ultimately, this produces political power (these companies not only spend significant sums lobbying policymakers but try to cultivate a positive public image) and shapes the research ecosystem. They fund far more AI research than governments or philanthropic foundations, so the resulting technologies are likely to reflect their needs and priorities. Even most (58 percent) of "ethical tech" researchers receive their funding from industry,[45] which likely limits the strength of their critique.

Finally, AI researchers–whether in industry or academia–are demo-graphically homogeneous. In the United States, most of the people with an undergraduate degree in computer science are male and either white or Asian.[46] Likely as a result, the industry is less diverse than the private sec-tor as a whole.[47] The demographic homogeneity also creates an alienating masculine culture in innovation spaces, which then reproduces the prob-lem.[48] Employees from disadvantaged communities of color are more likely to provide catering or custodial services.[49] The global landscape echoes these inequalities, with workers receiving shockingly low wages to perform high-stress jobs like content moderation and image tagging.[50]

Structural inequality has even deeper roots in the history of science and technology. The power of innovators with formal technical expertise, who can contribute to the marketplace, has erased the contributions of indige-nous knowledge systems until relatively recently,[51] not to mention the ex-periential knowledge of citizens who may have different priorities than scientists or engineers. A long legacy of mistreating the participants in research—including the famous Tuskegee syphilis experiment, the use of Henrietta Lacks's cells across biomedicine, and the unhoused Black men who did not consent to improve Google's facial recognition technology—has led marginalized communities to be skeptical of technological innova-tion even when it is designed to benefit them. Significant portions of the US Black community, for example, have refused the COVID-19 vaccine because they question the intense public health attention. Marginalized communities have also experienced devastating neglect, which can be a matter of life and death. For years, scientists have known that the pulse ox-imeter, used to measure blood oxygen during the COVID-19 pandemic, was inaccurate among those with darker skin. As a result, those who need supplemental oxygen may not receive it.[52] It was only after 2020, when an anthropologist sounded the alarm in the wake of George Floyd's murder and physicians confirmed the problem, that health ministries around the world took notice.[53] While technical communities and policymakers may treat such problems as minor, isolated errors, marginalized communities see them as examples of structural inequality, justifying their frustration with science and technology.

Ultimately, the concentration of power in the tech industry combined with structural inequality make it very difficult to produce more equitable and just AI. A handful of tech leaders shape the definition of AI problems and promote a simplistic understanding of the relationships between technology and society, including assumptions that technologies usually have beneficial impacts and can easily fix societal ills. These are the understandable assumptions of those whose lives have generally improved with technology, but they have serious consequences for others.

Bringing Power into AI Equity Solutions

To ensure that AI ameliorates, or at least does not exacerbate, the structural inequities we have identified, we must reimagine the four types of solutions already described. Technical solutions that account for power would focus scientists and engineers on the concerns of marginalized people, rather than the other way around.[54] This starts with agenda-setting: Research funders or technologists might begin by asking a community about the biggest challenges they face and then determine development priorities accordingly. The partnership would continue throughout the design process, so that citizens may provide their expertise and feel some ownership over the project and so researchers can establish trust with the community.

In Pittsburgh, for example, a technical team led by computer scientists at Carnegie Mellon University (CMU) worked with community members to build a technology that monitored and visualized local air quality.[55] The collaboration began when the researchers attended community meetings and learned about residents' concerns about air pollution from a nearby factory. Residents had previously struggled to get the attention of local or national officials because they were unable to produce enough quantitative data in a timely fashion. The researchers listened to the residents' plight, built prototypes, and then altered the technology in response to community input. Eventually, their system brought together heterogeneous data, including crowdsourced smell reports, animated smoke images, finer air quality data, and wind information, which the community then used to trigger government action—EPA administrators agreed to review the factory's compliance, and later that year, the parent company announced its closure. This approach, however, required openness and humility from the researchers, recognition of community expertise, a desire to empower marginalized people, and willingness to suppress technical priorities in favor of the needs of the neighborhood.[56]

Organizational solutions that alleviate structural inequality require leaders to identify how culture, language, norms, and daily practices can reinforce the power of certain groups and then work to change them. Diversifying an organization without this attention will simply produce more alienation and scandal. Technical organizations must clearly demonstrate their openness to hear hard truths about their own privilege, understand how historically disadvantaged people may be disproportionately harmed by their work, and prioritize solutions. To achieve this, all tech companies should have teams that focus on the equity dimensions of AI and report directly to the CEO. Such teams would weigh in on major research and development decisions, would be given long-term funding commitments, and would receive whistleblower protections.[57] Universities also have an important role to play in training the next generations of scientists and engineers to understand the discrimination and harms perpetrated by their forebears; few people know, for example, that the academic field of statistics—which underlies AI—is rooted in eugenic ideology.[58] Today, universities may require STEM students to take a single course on professional ethics.[59] Instead, they should integrate attention to the equity, social, and ethical impacts of AI into core technical courses.[60] And humanists and social scientists should teach this content to disrupt the conventional privileges afforded technical experts. After all, these experts offer deep knowledge of how technology works *in society.* Finally, government agencies and philanthropic foundations who have begun to encourage research into the implications of AI should facilitate equitable multidisciplinary collaborations.

Scholars have envisioned a variety of legal and policy tools that take power imbalances seriously. This includes algorithmic impact assessments (AIAs), which governments could use to assess the risks and benefits of a particular technology before it is deployed.[61] Similar to environmental impact assessments required for new development projects and government reviews of new drugs, they would require government officials to answer a standard battery of questions about the impacts of the system's technical attributes, which would result in a final impact score that would determine its regulation.[62] However, focusing on the technical dimensions of the system is insufficient. AIAs must consider the social implications. In its report on the benefits and harms of facial recognition technology in K-12 schools, New York state's Office of Information Technology Services considered not only accuracy but also the likelihood that the technology would exacerbate bias and harm against already marginalized communities.[63] Even if facial recognition technology became more accurate, the Office concluded, it would violate civil rights and liberties. The state legislature banned this use in response to the report.[64]

Others have suggested more deliberative approaches to increase civic capacity. In the case of genomics and biotechnology, Osagie Obasogie advocates for race impact assessments that are collaborative and involve multiple stakeholders.[65] Systematically incorporating marginalized communities into algorithmic impact assessments could also help to empower them and ultimately alleviate structural inequalities. Key, however, is to link democratic deliberation to decision-making; otherwise these citizens will feel further exploited and neglected.[66] Before launching an initiative to bring people with disabilities more centrally into tech innovation, for example, Borealis Philanthropy and the Ford Foundation appealed to an advisory committee made up of people with disabilities who offered a range of expertise.[67] Over the course of a year, the committee identified priorities, offered strategies to address harms at the intersection of disability and technology, and nominated and selected the inaugural cohort of grantees. Experts can also help community organizations advocate for more just AI development and use. The University of Michigan's Science, Technology, and Public Policy Program has established the Community Partnerships Initiative, which responds to the concerns and priorities of organizations in Detroit and southeast Michigan with research and analysis.[68] For example, it produced a policy brief on acoustic gun detection systems, which enabled We the People Michigan to challenge Detroit's investment in the technology.[69]

Distributing Responsibility for AI and Equity

More serious attention to structural inequality and the power imbalances they produce will require all the participants in the innovation ecosystem—innovators, customers, funders, regulators, and the public—to take on additional responsibilities. AI innovators must abandon the notion that their work is politically neutral and objective and recognize that if they seek societal benefit rather than harm, they must engage a diverse populace throughout the process, even at the priority-setting stage. They must treat these communities with respect, which includes taking their advice especially when they sound alarms, paying them, and making transparent decisions.[70] Innovators must also understand that social context will shape the impact of technologies they build, both positively and negatively. In other words, technologies are only solutions if they fit with the culture, conventions, and relationships in a particular place. For interventions to have the benefits technologists seek, they should work with historians, sociologists, and anthropologists who can offer deep understanding of communities and the relationships between technology and society.

Meanwhile, those that purchase AI must develop the capacity to inquire about datasets and algorithms and the structural inequalities that they may hide and perpetuate. In some cases, they may be able to force technologists to change the technology. But even when they cannot, they can guide those who ultimately use the technology regarding its limitations and processes that may minimize harm to vulnerable communities.

Funders, whether public, philanthropic, or private, also have an important role to play. They can include marginalized communities on advisory committees that set funding priorities, and privilege these insights with the understanding that they have had virtually no voice in the history of technological innovation thus far.[71] Funders will also need to think quite differently about innovation. For AI to achieve important goals such as improving cancer survival rates or mitigating climate change among vulnerable communities, funders must recognize that the problems are simultaneously social and technical and create research opportunities accordingly. Funders can provide incentives, or even require technologists to collaborate with marginalized communities, humanists, and social scientists on individual projects, to ensure that they redress historical inequities. Private sector funders can provide incentives to technologists who consider equity and justice explicitly in their work; these developments will likely open new markets, which will ultimately benefit investors as well.

Regulators around the world have begun to take some responsibility for AI. In the United States, the Biden administration's recent Executive Order aims to provide guidance on responsible use to the users of algorithms across multiple sectors, including housing, criminal justice, and benefits programs with guidance on responsible use.[72] It is also developing systems to evaluate AI safety including requiring developers to disclose the results of their "red team" tests. But this is not enough and is likely to focus regulators on the technical dimensions of the systems.[73] We suggest that regulators consider more comprehensive impact assessments. This would require not only technical investigation of the datasets and algorithms but also the consequences when the technology is deployed in society. In other words, regulators will have to move beyond technical evaluation, which will require them to incorporate new types of expertise and evaluation processes.

As innovators, customers, regulators, and funders take on these new responsibilities, it will place new burdens on already marginalized communities. Their inclusion is crucial to achieve equity and justice, but it is also risky. They may be overwhelmed by requests, tokenized, or provided insufficient compensation for their participation. They may also simply be wary of being ignored or abused, given the history of their participation in innovation. They must always have the agency to say no, and the innovation ecosystem must accept this. When they choose to participate, their knowledge must be valued and compensated fairly. This is the only way to build trust and ultimately alleviate structural inequities in AI and innovation more generally.

Notes

1. ACLU of Michigan, "After Third Wrongful Arrest, ACLU Slams Detroit Police Department for Continuing to Use Faulty Facial Recognition Technology," American Civil Liberties Union, August 6, 2023, https://www.aclu.org/press-releases/after-third-wrongful-arrest-aclu-slams-detroit-police-department-for-continuing-to-use-faulty-facial-recognition-technology.

2. Rashida Richardson, Jason M. Schultz, and Kate Crawford, "Dirty Data, Bad Predictions: How Civil Rights Violations Impact Police Data, Predictive Policing Systems, and Justice," *New York University Law Review Online* 94 (2019): 15–55.

3. Bobby Allyn, "In Settlement, Facebook to Pay $52 Million to Content Moderators with PTSD," NPR, May 12, 2020, sec. Technology, https://www.npr.org/2020/05/12/854998616/in-settlement-facebook-to-pay-52-million-to-content-moderators-with-ptsd; Billy Perrigo, "Exclusive: OpenAI Used Kenyan Workers on Less Than $2 Per Hour to Make ChatGPT Less Toxic," *TIME*, January 18, 2023, https://time.com/6247678/openai-chatgpt-kenya-workers/; Miriah Steiger et al., "The Psychological Well-Being of Content Moderators: The Emotional Labor of Commercial Moderation and Avenues for Improving Support," in *Proceedings of the 2021 CHI Conference on Human Factors in Computing Systems*, 2021 CHI Conference on Human Factors in Computing Systems (New York: Association for Computing Machinery, 2021), 1–14, https://doi.org/10.1145/3411764.3445092.

4. Todd Feathers, "People with Disabilities Say This AI Tool Is Making the Web Worse for Them," *Vice*, March 17, 2021, https://www.vice.com/en/article/m7az74/people-with-disabilities-say-this-ai-tool-is-making-the-web-worse-for-them.

5. Office of Science and Technology Policy, "Blueprint for an AI Bill of Rights," OSTP, The White House, 2022, https://www.whitehouse.gov/ostp/ai-bill-of-rights/.

6. Andrew Tutt, "An FDA for Algorithms," *Administrative Law Review* 69, no. 1 (2017): 83–123; Ryan Calo, "The Case for a Federal Robotics Commission," Brookings, September 15, 2014, https://www.brookings.edu/articles/the-case-for-a-federal-robotics-commission/.

7. Sanders Kleinfeld, "A New Course to Teach People about Fairness in Machine Learning," *The Keyword, Google* (blog), October 18, 2018, https://blog.google/technology/ai/new-course-teach-people-about-fairness-machine-learning/; Brianna Richardson and Juan E. Gilbert, "A Framework for Fairness: A Systematic Review of Existing Fair AI Solutions," *arXiv*, December 10, 2021, https://doi.org/10.48550/arXiv.2112.05700.

8. Shobita Parthasarathy, "Innovating for Equity," *Issues in Science and Technology* 38, no. 3 (Spring 2022), https://issues.org/innovating-for-equity-shobita-parthasarathy-forum/.

9. National Academies of Sciences, Engineering, and Medicine and National Academy of Medicine, *Toward Equitable Innovation in Health and Medicine: A Framework* (Washington, DC: National Academies Press, 2023), https://doi.org/10.17226/27184.

10. Tawanna R. Dillahunt and Tiffany C. Veinot, "Getting There: Barriers and Facilitators to Transportation Access in Underserved Communities," *ACM Transactions on Computer-Human Interaction* 25, no. 5 (October 11, 2018): 1–39, https://doi.org/10.1145/3233985.

11. Sasha Costanza-Chock, *Design Justice: Community-Led Practices to Build the Worlds We Need* (Cambridge, MA: MIT Press, 2020), https://doi.org/10.7551/mitpress/12255.001.0001.

12. Solon Barocas, Moritz Hardt, and Arvind Narayanan, *Fairness and Machine Learning: Limitations and Opportunities* (Cambridge, MA: MIT Press, 2023).

13. Algernon Austin, "My Phone's Facial Recognition Technology Doesn't See Me, a Black Man. But It Gets Worse," *USA Today*, December 17, 2019, https://www.usatoday.com/story/opinion/voices/2019/12/17/artificial-intelligence-facial-recognition-technology-black-african-american-column/2664575001/.

14. Max Plenke, "The Reason This 'Racist Soap Dispenser' Doesn't Work on Black Skin," *Mic*, September 9, 2015, https://www.mic.com/articles/124899/the-reason-this-racist-soap-dispenser-doesn-t-work-on-black-skin.

15. Joy Buolamwini and Timnit Gebru, "Gender Shades: Intersectional Accuracy Disparities in Commercial Gender Classification," in *Proceedings of the 1st Conference on Fairness, Accountability and Transparency*, Conference on Fairness, Accountability and Transparency, PMLR (2018), 77–91, https://proceedings.mlr.press/v81/buolamwini18a.html.

16. Inioluwa Deborah Raji and Joy Buolamwini, "Actionable Auditing: Investigating the Impact of Publicly Naming Biased Performance Results of Commercial AI Products," in *Proceedings of the 2019 AAAI/ACM Conference on AI, Ethics, and Society*, AIES '19 (New York: Association for Computing Machinery, 2019), 429–435, https://doi.org/10.1145/3306618 .3314244.

17. Mona Sloane et al., "Participation Is Not a Design Fix for Machine Learning," in *Proceedings of the 2nd ACM Conference on Equity and Access in Algorithms, Mechanisms, and Optimization*, EAAMO '22 (New York: Association for Computing Machinery, 2022), 1–6, https://doi.org/10.1145/3551624.3555285.

18. Sarah Brayne, *Predict and Surveil: Data, Discretion, and the Future of Policing* (New York: Oxford University Press, 2020).

19. Jennifer Elias, "Google Contractor Reportedly Tricked Homeless People into Face Scans," *CNBC*, October 3, 2019, sec. Technology, https://www.cnbc.com/2019/10/03/google -contractor-reportedly-tricked-homeless-people-into-face-scans.html.

20. Maggie Harrison Dupré, "Facial Recognition Used to Evict Single Mother for Taking Night Classes," *Futurism* (blog), May 17, 2023, https://futurism.com/facial-recognition -housing-projects.

21. Microsoft, "Responsible AI Standard, V2," June 2022, https://blogs.microsoft.com/wp -content/uploads/prod/sites/5/2022/06/Microsoft-Responsible-AI-Standard-v2-General -Requirements-3.pdf.

22. Mozilla, "$2.4 Million in Prizes for Schools Teaching Ethics Alongside Computer Science," *Distilled—The Mozilla Blog* (blog), April 30, 2019, https://blog.mozilla.org/en /mozilla/2-4-million-in-prizes-for-schools-teaching-ethics-alongside-computer-science/.

23. Madhumita Murgia and Cristina Criddle, "Big Tech Companies Cut AI Ethics Staff, Raising Safety Concerns," *Financial Times*, March 29, 2023, sec. Artificial intelligence, https:// www.ft.com/content/26372287-6fb3-457b-9e9c-f722027f36b3.

24. Katharine Miller, "Ethics Teams in Tech Are Stymied by Lack of Support," *Human-Centered Artificial Intelligence, Stanford University* (blog), June 21, 2023, https://hai.stanford .edu/news/ethics-teams-tech-are-stymied-lack-support; Emanuel Moss and Jacob Metcalf, "Ethics Owners: A New Model of Organizational Responsibility in Data-Driven Technology Companies," *Data & Society*, September 23, 2020, https://datasociety.net/library/ethics -owners/.

25. Emily M. Bender et al., "On the Dangers of Stochastic Parrots: Can Language Models Be Too Big? 🦜," in *Proceedings of the 2021 ACM Conference on Fairness, Accountability, and Transparency*, FAccT '21 (New York: Association for Computing Machinery, 2021), 610–623, https://doi.org/10.1145/3442188.3445922.

26. Tom Simonite, "What Really Happened When Google Ousted Timnit Gebru," *WIRED*, June 8, 2021, https://www.wired.com/story/google-timnit-gebru-ai-what-really -happened/.

27. Johanna Okerlund et al., "What's in the Chatterbox? Large Language Models, Why They Matter, and What We Should Do About Them," Science, Technology and Public Policy

(STPP), Gerald R. Ford School of Public Policy, University of Michigan, April 2022, https://stpp.fordschool.umich.edu/research/research-report/whats-in-the-chatterbox.

28. Tamara Kneese, "Climate Justice & Labor Rights," *AI Now*, August 2, 2023, https://ainowinstitute.org/general/climate-justice-and-labor-rights-part-i-ai-supply-chains-and-workflows.

29. Elizabeth Dwoskin and Nitasha Tiku, "A Recruiter Joined Facebook to Help It Meet Its Diversity Targets. He Says Its Hiring Practices Hurt People of Color," *Washington Post*, April 9, 2021, https://www.washingtonpost.com/technology/2021/04/06/facebook-discrimination-hiring-bias/.

30. "Google Gives Black Workers Lower-Level Jobs and Pays Them Less, Suit Claims," *The Guardian*, March 18, 2022, sec. Technology, https://www.theguardian.com/technology/2022/mar/18/google-black-employees-lawsuit-racial-bias.

31. *Facial Recognition Services: Federal Law Enforcement Agencies Should Take Actions to Implement Training, and Policies for Civil Liberties* (Washington, DC: US Government Accountability Office, September 12, 2023), https://www.gao.gov/products/gao-23-105607.

32. Ivana Saric, "NYC Law Promises to Regulate AI in Hiring, but Leaves Crucial Gaps," *Axios*, July 6, 2023, https://www.axios.com/2023/07/06/new-york-ai-hiring-law.

33. Ellen P. Goodman and Julia Trehu, "AI Audit Washing and Accountability," SSRN Scholarly Paper (Rochester, NY: September 22, 2022), https://doi.org/10.2139/ssrn.4227350.

34. Ally Jarmanning, "Boston Lawmakers Vote to Ban Use of Facial Recognition Technology by the City," NPR, June 24, 2020, sec. America Reckons with Racial Injustice, https://www.npr.org/sections/live-updates-protests-for-racial-justice/2020/06/24/883107627/boston-lawmakers-vote-to-ban-use-of-facial-recognition-technology-by-the-city.

35. Aidan Peppin, "Listening to the Public," Ada Lovelace Institute, August 18, 2023, https://www.adalovelaceinstitute.org/report/listening-to-the-public/; Octavia Reeve, Anna Colom, and Roshni Modhvadia, "What Do the Public Think about AI?," Ada Lovelace Institute, October 26, 2023, https://www.adalovelaceinstitute.org/evidence-review/what-do-the-public-think-about-ai/.

36. "NSF and Partners Kick off the National Artificial Intelligence Research Resource Pilot Program," US National Science Foundation, November 9, 2023, https://new.nsf.gov/news/nsf-partners-kick-nairr-pilot-program.

37. Nur Ahmed, Muntasir Wahed, and Neil C. Thompson, "The Growing Influence of Industry in AI Research," *Science* 379, no. 6635 (March 3, 2023): 884–886, https://doi.org/10.1126/science.ade2420.

38. Alex Hanna and Emily M. Bender, "AI Causes Real Harm. Let's Focus on That over the End-of-Humanity Hype," *Scientific American*, August 12, 2023, https://www.scientificamerican.com/article/we-need-to-focus-on-ais-real-harms-not-imaginary-existential-risks/; Lorena O'Neil, "These Women Tried to Warn Us About AI," *Rolling Stone*, August 12, 2023, https://www.rollingstone.com/culture/culture-features/women-warnings-ai-danger-risk-before-chatgpt-1234804367/.

39. Will Knight, "A Letter Prompted Talk of AI Doomsday. Many Who Signed Weren't Actually AI Doomers," *Wired*, August 17, 2023, https://www.wired.com/story/letter-prompted-talk-of-ai-doomsday-many-who-signed-werent-actually-doomers/.

40. Mario Biagioli, "Patent Specification and Political Representation: How Patents Became Rights," in *Making and Unmaking Intellectual Property: Creative Production in Legal*

and Cultural Perspective, ed. Mario Biagioli, Peter Jaszi, and Martha Woodmansee (Chicago: University of Chicago Press, 2011), 25–40, https://press.uchicago.edu/ucp/books/book /chicago/M/bo11103013.html.

41. Stuart W. Leslie, *The Cold War and American Science: The Military-Industrial-Academic Complex at MIT and Stanford* (New York: Columbia University Press, 1993).

42. Sebastian M. Pfotenhauer, Joakim Juhl, and Erik Aarden, "Challenging the 'Deficit Model' of Innovation: Framing Policy Issues Under the Innovation Imperative," *Research Policy*, New Frontiers in Science, Technology and Innovation Research from SPRU's 50th Anniversary Conference, 48, no. 4 (May 1, 2019): 895–904, https://doi.org/10.1016/j.respol.2018 .10.015.

43. Roger McNamee, *Zucked: Waking Up to the Facebook Catastrophe* (New York: Penguin Press, 2020).

44. Chase Peterson-Withorn, "The 25 Richest People in the World 2023," *Forbes*, April 4, 2023, https://www.forbes.com/sites/chasewithorn/2023/04/04/the-25-richest-people-in-the -world-2023/; Felix Richter, "Tech Giants Do Heavy Lifting in 2023 Stock Market Rebound," *Statista Daily Data* (blog), June 19, 2023, https://www.statista.com/chart/30219/main -contributors-to-s-p-500-gains-in-2023.

45. Mohamed Abdalla and Moustafa Abdalla, "The Grey Hoodie Project: Big Tobacco, Big Tech, and the Threat on Academic Integrity," in *Proceedings of the 2021 AAAI/ ACM Conference on AI, Ethics, and Society*, AIES '21 (New York: Association for Computing Machinery, 2021), 287–297, https://doi.org/10.1145/3461702.3462563.

46. Shana Lynch, "2023 State of AI in 14 Charts," *Human-Centered Artificial Intelligence, Stanford University* (blog), April 3, 2023, https://hai.stanford.edu/news/2023-state-ai-14 -charts.

47. Susan Laborde, "30+ Diversity in High Tech Statistics [2023 Data]," *Tech Report*, May 28, 2024, https://techreport.com/statistics/business-workplace/diversity-in-high-tech -statistics/.

48. J. Lewis, "Barriers to Women's Involvement in Hackspaces and Makerspaces," Monograph (University of Sheffield, September 3, 2019), https://eprints.whiterose.ac.uk/144264/.

49. Jessica Guynn, "Race and Class Divide: Black and Hispanic Service Workers Are Tech's Growing Underclass," *USA Today*, July 10, 2020, sec. Tech, https://www.usatoday.com /story/tech/2020/07/10/black-hispanic-workers-tech-underclass-amazon-apple-facebook -google/13461027/.

50. Perrigo, "OpenAI Used Kenyan Workers on Less Than $2 Per Hour"; Billy Perrigo, "Inside Facebook's African Sweatshop," *TIME*, February 14, 2022, https://time.com/6147458 /facebook-africa-content-moderation-employee-treatment/.

51. Christopher I. Roos et al., "Native American Fire Management at an Ancient Wildland–Urban Interface in the Southwest United States," *Proceedings of the National Academy of Sciences* 118, no. 4 (January 26, 2021): e2018733118, https://doi.org/10.1073/pnas .2018733118; Charles R Menzies and Caroline F Butler, "Returning to Selective Fishing Through Indigenous Fisheries Knowledge: The Example of K'moda, Gitxaala Territory," *American Indian Quarterly* 31, no. 3 (2007): 441–464.

52. Amy Moran-Thomas, "How a Popular Medical Device Encodes Racial Bias," *Boston Review*, August 5, 2020, https://www.bostonreview.net/articles/amy-moran-thomas-pulse -oximeter/.

53. Center for Devices and Radiological Health, "Pulse Oximeter Accuracy and Limitations: FDA Safety Communication," US Food and Drug Administration, September 15, 2022, https://public4.pagefreezer.com/content/FDA/20-02-2024T15:13/https://www.fda.gov/medical-devices/safety-communications/pulse-oximeter-accuracy-and-limitations-fda-safety-communication.

54. Abeba Birhane, "Algorithmic Injustice: A Relational Ethics Approach," *Patterns* 2, no. 2 (February 12, 2021): 100205, https://doi.org/10.1016/j.patter.2021.100205.

55. Yen-Chia Hsu et al., "Community-Empowered Air Quality Monitoring System," *arXiv* April 9, 2018, https://doi.org/10.48550/arXiv.1804.03293.

56. Yen-Chia Hsu et al., "Empowering Local Communities Using Artificial Intelligence," *Patterns* 3, no. 3 (March 11, 2022): 100449, https://doi.org/10.1016/j.patter.2022.100449.

57. Christina J. Colclough and Kate Lappin, "Building Union Power to Rein in the AI Boss," *Stanford Social Innovation Review*, September 20, 2023, https://ssir.org/articles/entry/building_union_power_to_rein_in_the_ai_boss.

58. Ruth Schwartz Cowan, "Francis Galton's Statistical Ideas: The Influence of Eugenics," *Isis* 63, no. 4 (December 1972): 509–528, https://doi.org/10.1086/351000.

59. Sepehr Vakil, "Ethics, Identity, and Political Vision: Toward a Justice-Centered Approach to Equity in Computer Science Education," *Harvard Educational Review* 88, no. 1 (2018): 26–52, https://doi.org/10.17763/1943-5045-88.1.26.

60. James W. Malazita, "Translating Critical Design: Agonism in Engineering Education," *Design Issues* 34, no. 4 (October 1, 2018): 96–109, https://doi.org/10.1162/desi_a_00514; James W. Malazita and Korryn Resetar, "Infrastructures of Abstraction: How Computer Science Education Produces Anti-Political Subjects," *Digital Creativity* 30, no. 4 (October 2, 2019): 300–312, https://doi.org/10.1080/14626268.2019.1682616.

61. Emanuel Moss et al., "Assembling Accountability: Algorithmic Impact Assessment for the Public Interest," *Data & Society*, June 29, 2021, https://datasociety.net/library/assembling-accountability-algorithmic-impact-assessment-for-the-public-interest/.

62. Lara Groves, "Algorithmic Impact Assessment: A Case Study in Healthcare," Ada Lovelace Institute, February 8, 2022, https://www.adalovelaceinstitute.org/report/algorithmic-impact-assessment-case-study-healthcare/.

63. "Use of Biometric Identifying Technology in Schools," Office of Information Technology Services, New York State, August 2023, https://its.ny.gov/system/files/documents/2023/08/biometrics-report-final-2023.pdf.

64. Carolyn Thompson, "New York Bans Facial Recognition in Schools," *TIME*, September 27, 2023, https://time.com/6318033/new-york-bans-facial-recognition-schools/.

65. Osagie K. Obasogie, "Toward Race Impact Assessments," in *Beyond Bioethics: Toward a New Biopolitics*, ed. Osagie K. Obasogie and Marcy Darnovsky (Oakland, CA: University of California Press, 2018), 461–471.

66. Mark B. Brown, *Science in Democracy: Expertise, Institutions, and Representation* (Cambridge, MA: MIT Press, 2009).

67. "Borealis Philanthropy and Ford Foundation Launch $1 Million Disability x Tech Fund to Advance Leadership of People with Disabilities in Tech Innovation," *Ford Foundation* (blog), February 28, 2023, https://www.fordfoundation.org/news-and-stories/news-and-press/news/borealis-philanthropy-and-ford-foundation-launch-1-million-disability-x-tech-fund-to-advance-leadership-of-people-with-disabilities-in-tech-innovation/.

68. "Expanding Participation in Science and Technology Policy Through Civil Society Partnerships," *Gerald R. Ford School of Public Policy, University of Michigan* (blog), November 4, 2021, https://fordschool.umich.edu/news/2021/expanding-participation-science-and-technology-policy-through-civil-society-partnerships.

69. Jillian Mammino, "Acoustic Gunshot Detection Systems: Community & Policy Considerations" (Gerald R. Ford School of Public Policy, University of Michigan, June 2022).

70. "Community Engagement Playbook," Gerald R. Ford School of Public Policy, University of Michigan, forthcoming; "STPP to Explore Best Practice in Community Engagement," *Gerald R. Ford School of Public Policy, University of Michigan* (blog), August 1, 2023, https://fordschool.umich.edu/news/2023/stpp-explore-best-practice-community-engagement?theme=ipc.

71. Shobita Parthasarathy, "Can Innovation Serve the Public Good?," *Boston Review*, July 6, 2023, https://www.bostonreview.net/articles/can-innovation-serve-the-public-good/.

72. "Executive Order on the Safe, Secure, and Trustworthy Development and Use of Artificial Intelligence," The White House, October 30, 2023, https://www.whitehouse.gov/briefing-room/presidential-actions/2023/10/30/executive-order-on-the-safe-secure-and-trustworthy-development-and-use-of-artificial-intelligence/.

73. Sorelle Friedler et al., "AI Red-Teaming Is Not a One-Stop Solution to AI Harms: Recommendations for Using Red-Teaming for AI Accountability," *Data & Society*, October 25, 2023, https://datasociety.net/library/ai-red-teaming-is-not-a-one-stop-solution-to-ai-harms-recommendations-for-using-red-teaming-for-ai-accountability/.

CHAPTER 8

Scientific Progress in Artificial Intelligence: History, Status, and Futures

Eric Horvitz and Tom M. Mitchell

Introduction and Background

Artificial Intelligence (AI) refers to a field of endeavor as well as a constellation of technologies. The Association for the Advancement of AI (AAAI) defines the field as pursuing "the scientific understanding of the mechanisms underlying thought and intelligent behavior and their embodiment in machines." AI encompasses the development of methods for learning from data, representing knowledge, and performing reasoning aimed at building computer systems capable of performing tasks that typically have required human intelligence. Core capabilities covered in AI research include methods for learning, reasoning, problem-solving, planning, language understanding, and visual perception. Over the last twenty years, AI research transitioned from a niche scientific endeavor to an impactful set of technologies. We provide in this overview chapter a brief history of the evolution of AI as a discipline over nearly seven decades. Then, we review recent advances and directions. This arc through history, present, and the expected near future was commissioned to provide a February 2024 snapshot of the state of AI in support of a series of meetings on AI and the sciences that was organized by the National Academy of Sciences and the Annenberg Trust.

Birth and Evolution of Scientific Field

The prospect of automating aspects of human thinking via mechanical systems has been considered for hundreds of years. Modern metaphors and framing of thinking as a computational process have roots in the early twentieth century. Key contributions to the perspective of thinking as computing include the theoretical work of Alan Turing on computability,[1] efforts by John von Neuman, Turing, and others to construct general-purpose computing systems,[2] and work on computational abstractions of neuronal systems by McCollough and Pitts.[3] The 1940s saw the rise of discussions and publications viewing the computer as a metaphor for the brain, including control-theoretic notions referred to as *cybernetics*.[4]

The modern discipline of AI, per the establishment of a long-standing set of aspirations, harkens back to a research project proposal for a summer workshop held at Dartmouth College in 1956.[5] The proposal, coauthored by John McCarthy, Marvin Minsky, Nathaniel Rochester, and Claude Shannon, outlined a new field of studying how machines could be programmed to perform "every aspect of learning or any other feature of intelligence." Containing the first use of the phrase *artificial intelligence*, the proposal described goals of finding "how to make machines use language, form abstractions and concepts, solve kinds of problems now reserved for humans, and improve themselves." The summer study is considered as the formal launch of AI as a distinct field of scientific inquiry, setting the foundation for decades of research in computer science.

The maturation of the AI research program saw the evolution of a set of AI subdisciplines with overlapping but distinct research communities, including natural language understanding, problem-solving, planning, vision, robotics, and machine learning. Research areas and communities also formed around distinct foundational approaches to building AI, such as logical reasoning and representations, reasoning under uncertainty with statistical methods, and the use of neural network models versus high-level symbols—a domain of research that had been referred to for decades as *connectionist* approaches. Further, advances and questions in AI have stimulated efforts in other disciplines, such as cognitive psychology, where *cognitive science* refers to a subdiscipline of both AI and cognitive psychology that centers on taking inspiration from studies, data, and questions about human cognition to build systems that can perform automated learning and reasoning, and on using computational approaches to modeling and probing human psychological processes.[6]

Representations and Reasoning Mechanisms

Scientific studies of AI are best understood in terms of the technical evolution of different approaches to representing and reasoning with data and knowledge. In the early days of the field, representations and reasoning methods included the use of neural networks, early-on referred to as *perceptrons* in work on learning to recognize visual patterns,[7] and symbolic logic applied in both specific instances and in attempts to build general architectures for problem-solving.[8] Symbolic representations dominated the first several decades of AI research with efforts in statistical methods, including neural networks, continuing but largely taking a backstage position. Work in logic-based systems included *rule-based expert systems* that came to focus of attention in the 1970s and 1980s. These systems were aimed at capturing specialist knowledge in sets of compact logical rules (e.g., if-then rules) that would be used to compose chains of inferences within an architecture referred to as a *production system.*[9]

In a paradigm shift in the mid-1980s, attention began to shift from logic-based methods to statistical approaches for handling uncertainties associated with the complexity of real-world problems, such as applications in medical diagnosis and decision support. Representation and reasoning machinery were developed for harnessing probability theory and decision theory,[10] including *Bayesian networks*[11] and, more generally, *probabilistic graphical models.*[12] Systems were developed using these probabilistic representations for making inferences, such as inferring medical diagnoses from information about a patient's illness, sets of symptoms, and lab results. In some systems, the collection of additional information to help refine conclusions or diagnoses was guided by computing the *expected value of information* of additional observations, tests, or data.[13] In addition, AI research scientists began to incorporate and extend techniques developed in the related disciplines of Operations Research, such as *Markov decision processes* to support sequential decisions.[14]

Despite the rise and fall of excitement in different methods, efforts have continued within and across multiple fundamental representation and reasoning methods. For example, today's successes and focus of attention on large-scale neural networks extends in a recognizable line from the nascent work in the early 1960s on *perceptrons* to the most recent developments with methods and systems based on neural networks. Today, studies of symbolic reasoning methods continue, including on mechanisms for integrating symbolic reasoning with neural models to bolster their abilities to perform logic and more general mathematics.[15]

Machine Learning: Foundation of Today's AI

Machine learning involves algorithms that enable computers to automatically improve their performance at some task through experience. Often that experience takes the form of a large dataset (e.g., in systems that learn to classify which new credit card transactions are likely to be legitimate versus fraudulent) by training on large historical datasets of transactions where the correct classification is known in retrospect. In other cases, training experience may involve active experimentation, as in AI systems that learn to play games by using their evolving current best strategy to play against itself, to collect data on which game moves produce a win. Breakthroughs in AI over the last fifteen years are largely attributable to advances in machine learning. Today, machine learning is viewed as foundational to the field as AI moves into the future.

Beyond the aforementioned early research with perceptrons, today's scientific studies of machine learning extend back to numerous early efforts with learning from data or experience. Such efforts include game-playing systems in chess and checkers and research efforts that laid out surprisingly modern sets of concepts, flows, and architectures for machine learning.[16] For example, research on the Pandemonium system by Oliver Selfridge called out principles of salient feature discovery and the use of multiple levels of representation.[17]

Machine learning research accelerated in the late-1990s. During that time, algorithmic advances, construction of prototypes, and undertaking of empirical studies were catalyzed by the fast-paced rise in computing power and data storage capabilities, along with the explosion in the quantity of online data available for research and development. In the mid-1990s, large amounts of data started to become available via precipitous drops in cost of storage, new data capture technologies, and the massive quantity of content and behavioral data coming with the growth of the web.

A tapestry of machine learning methods has been developed over the last thirty years, many extending methods in traditional statistical analyses to handling datasets with larger numbers of variables and cases and frequently aimed at solving aspirational goals of AI. Enabling advances include methods developed in the late 1980s and early 1990s for directly learning probabilistic graphical models from data[18] and for enhancing the efficiency and capabilities of neural network constructions.[19]

Particularly important to where we are today with the science of AI—and powering the fast-paced progress in research and development—are scientific advances with harnessing multilayered neural networks that came to be referred to as a methodology named *deep learning*. Advances in deep learning have propelled AI to unprecedented levels of capabilities and utility. Innovations, stemming back decades, include the method of *back-propagation* for tuning multilayered neural networks with data[20] and *convolution*,[21] an approach to pooling complex signals into higher-level abstractions.

Discriminative and Generative Models

Machine learning methods can be broadly divided into two main capabilities, *discriminative AI* and *generative models*, each with distinct objectives and application categories. Discriminative models take as input the description of some item and outputs a label, or classification, of the item. For example, in the case of a junk email filter, a discriminative model learns to label each input email as either spam or non-spam by analyzing features derived from the email. These discriminative models directly use the features of the input data to make predictions or classifications, focusing on the relationship between the input data and its corresponding labels.

Discriminative models span classic statistical models of logistic regression, algorithms for learning classifiers from tabular data, and deep learning for diagnosis and classification. Examples of discriminative models include leveraging labeled data drawn from electronic health record systems to predict readmission,[22] sepsis,[23] and the onset of infection[24] in hospitalized patients.

Generative models have been front and center in the recent excitement about AI and its applications. Such models replicate the process by which data is generated. By learning the probability distribution of output features given input features, generative models can create and output new data instances that resemble the training data (in contrast to the labels output by discriminative models). Multiple methods have been used in generative AI, including techniques named *generative adversarial networks* (GANs), *variational autoencoders* (VAEs), *diffusion modeling*, and more recently, *transformers* that yield exciting capabilities of generative AI models. Generative models trained on images are now being used to generate novel imagery, as has become popularized in the DALL-E and Midjourney applications. Beyond images, generative methods are being used in a wide range of applications, including the structure and design of protein sequences and the performing of scientific simulations.

Supervised, Unsupervised, and Self-Supervised Learning

The training procedures by which models are constructed in machine learning can be broadly categorized into supervised, unsupervised, and self-supervised learning. *Supervised learning* relies on labeled datasets. The use of such curated data has been the basis of significant advancements in areas like medical diagnosis, image analysis, and speech recognition. *Unsupervised learning* refers to methods that find patterns in data without explicit labeling. Traditional variants of these methods include clustering and anomaly detection, which have been particularly useful in exploratory data analysis.

Over the last decade, a special form of *unsupervised learning*,[25] named *self-supervised learning*, has become very important. Self-supervision is a simple yet powerful idea that has enabled AI systems to learn from vast unlabeled datasets, such as massive corpora, crawled from across the web. One approach to self-supervision is to generate labels automatically by a "fill in the blanks" process of hiding words in text or other types of tokens in datasets and then trying to predict the hidden information. As an example, a model might predict the next word in a sentence or the next frame in a video sequence based on previous words or frames.

Self-supervised learning represents a significant shift in machine learning, moving away from heavy reliance on human-labeled data. This paradigm is unlocking new possibilities across various fields, enabling models to learn from vast untapped datasets and driving innovation in areas where labeled data is scarce or expensive to obtain.

Inflection Point for AI: Deep Learning

We are now experiencing an inflection in AI with an acceleration in the rate of innovation. The acceleration is largely attributable to advances in research and development with *deep neural networks* (DNNs) over the last decade.

Excitement about the potential of DNNs was sparked by surprising results in speech recognition, natural language processing, and machine vision. In 2009, DNN methods surprised the community with an unexpected reduction in word error recognition rates challenging conversational speech recognition tasks, including one named Switchboard.[26] Progress on the Switchboard benchmark had essentially plateaued for over a decade when progress was made with a DNN approach. Shortly after these gains in speech recognition, another DNN model named AlexNet was developed and demonstrated to perform with surprising capability on an object recognition challenge dataset named ImageNet.[27]

Since that time, research and applications with DNNs have exploded with new challenge problems and applications. Over the last five years, neural models have been used in multiple applications, including scene recognition systems used in semiautonomous driving. In another domain, DNNs

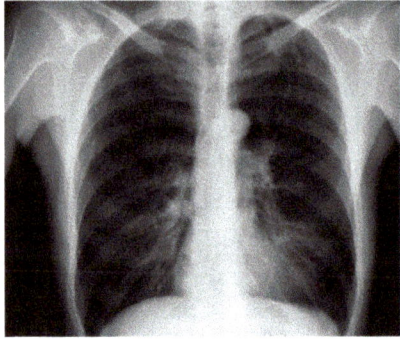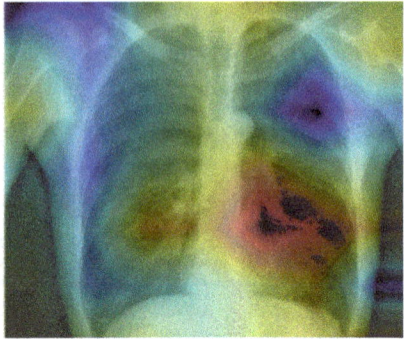

Figure 8.1. Visualization generated by CheXNet model, highlighting a region in a radiological image of the thorax, where the system recognizes right pleural diffusion. Pranav Rajpurkar et al., "CheXNet: Radiologist-Level Pneumonia Detection on Chest X-Rays with Deep Learning" *arXiv*, December 25, 2017, https://doi.org/10.48550/arXiv.1711.05225.

have been demonstrated to perform at expert levels with interpreting medical imagery. For example, DNNs have been shown to have the capability to provide expert-level classifications, such as the diagnosis of dermatological disorders from images of skin[28] and diagnoses from radiological films (Figure 8.1).[29]

Sets of evaluation benchmarks have been defined in the language and vision areas, such as the General Language Understanding Evaluation (GLUE), a benchmark formulated to measure the performance of models with language understanding across a range of natural language processing tasks.[30] In stunning advances over a decade, AI systems have reached parity with humans on numerous of the defined challenge problems, as highlighted in Figure 8.2. Details of the progress on the capabilities of AI systems has been captured in the recurrent reports of the AI Index, an annual study of trends in AI hosted at Stanford University.[31]

Key Concepts and Research Directions

Several key directions have come to the fore as important developments, requirements, and directions in work on DNNs, including interest in

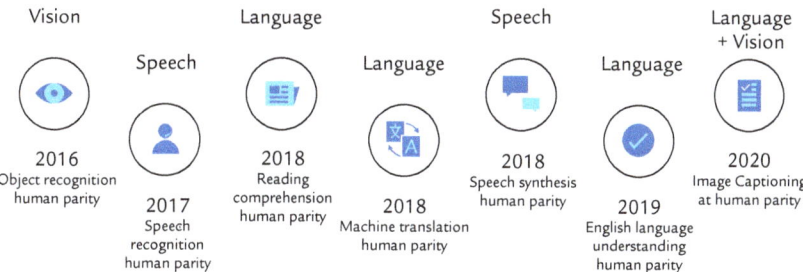

Figure 8.2. AI at an inflection. Deep neural networks have fueled advances in capabilities on benchmarks designed as top challenges for AI systems. This figure shows competencies of AI models where human parity on the challenge problems was reached on seven benchmarks.

automated ways to learn rich representations rather than curate them with expert guidance; efforts on robustness; the weaving together of multi-modal datasets; and the critical value of hardware, innovative algorithms, and programming platforms for research and development.

Learning representations. Early supervised machine learning required the identification of handcrafted salient observations or *features* of the input and based its predictions on those handcrafted features. Researchers have explored how deep learning can identify such features automatically or, more generally, rich representations directly from fine-grained data, in processes referred to as *representation learning.* The automated learning of rich features to represent an image, starting with the lowest-level input pixel features and becoming progressively more complex and abstract at successive layers of neural networks has been a celebrated aspect of modern deep neural models for vision.

The ability to automatically learn, or discover, candidate features enables systems to discover how best to organize the structure of machine learning problems, often yielding more accurate and robust performance on complex tasks than human-defined attributes, albeit at the cost of increased data requirements. Neural models leveraging such representation learning have been developed for natural language processing, computer vision, speech recognition, and health care.

Robustness and generalization. Efforts in the field of DNNs have increasingly focused on achieving robustness and generalization to ensure accurate performance in varying real-world environments that can be assumed to require robust capabilities, for example, accurate diagnoses, classifications, and predictions on new, previously unseen data, that is not contained in training datasets. Efforts in this realm push DNN training procedures to seek universal patterns from their training data so as to reduce their being overfit to training data and to be more adaptable to diverse real-world scenarios.

Hybrid strategies. Successes have been found with combining DNNs with other computational methods such as coupling the neural models with scientific simulations, integrating the methods with Markov decision processes (e.g., reinforcement learning), and integrating DNN approaches with symbolic approaches to reasoning. As an example, the AlphaGo systems rely on an integration of deep neural models for making predictions with reinforcement learning for guiding the choice of actions.[32]

Multimodal modals. Most DNN efforts have focused on the singular modalities of language or visual tasks. In the spirit of pursuing more human-like intelligence, researchers have pursued the development of multimodal models that bring together language, imagery, sounds, and other modalities. Multimodal DNNs include early efforts to do image captioning and more recent efforts to make inferences about language and images for such tasks as writing radiological reports.

Tools, methods, and platforms. With the advent of the importance of DNNs and growing focus of attention on using increasingly large datasets, methods have been pursued for introducing new forms of efficiency via hardware and algorithmic innovation, and for developing programming environments doing exploratory work with architectural designs for neural networks. At the hardware level, graphical processing units (GPUs) have provided speed-ups via parallel processing of matrix and vector operations that are central to deep learning.

Algorithmically, efforts span methods for introducing new forms of speed-ups in distributed computing at the hardware system level as well as on higher-level software innovations aimed at speeding-up the core back-propagation procedure to identify parameters that specify weights of connections in neural models. For example, efforts have focused on adaptation of mathematical optimization procedures like stochastic gradient descent.

Programming environments such as TensorFlow and PyTorch were created to ease the design and testing of DNNs, providing engineers with computing libraries, methods for accelerating GPU computation, and efficiently specifying and revising the structure of neural networks.

Models as platforms. For decades in machine learning, researchers have studied methods for adapting models trained on a source task to perform well on other domains via processes of *fine-tuning* the models with specialized data. This process, often referred to as *transfer learning*, leverages the knowledge that the model has gained from the initial training to perform well on a related, but different, task.[33] Large-scale neural models can serve as platforms for extending via fine-tuning with specialized datasets drawn from target task domains. Given the myriad uses of the large models as platforms that can be extended via domain-specific data, they have been referred to as foundation models.[34] Foundation models can be seen as an extension and scaling-up of transfer learning to DNNs that are trained on extremely large datasets, often encompassing a wide range of topics, languages, or modalities. Their versatility lies in the ability of pretrained models to be fine-tuned with smaller, task-specific datasets, thereby reducing the need for training a model from scratch for each new application. This approach not only saves significant computational resources but also allows for building upon the model's base capabilities and knowledge. The term *foundation* reflects their role as a fundamental base upon which more specialized or *fine-tuned* models can be built, similar to how a foundation supports a structure. Their general-purpose nature and scalability make them akin to a utility or resource that can be tapped into for numerous AI systems. Fine-tuning pretrained foundation models has become a standard methodology for adding new capabilities, such as adding multimodal capabilities to language-only models[35] and for extending the power of generalist models to specialist performance.[36]

A Second Inflection: Generative AI

The landscape of AI and its influences on the world has now reached a second inflection, *Generative AI*. Generative AI models are rich language and multimodal models that are trained to predict sequences of outputs, given input sequences or *prompts*. These generative models generate the output sequence one item at a time, at each step considering the newest generated item as a new part of the input, as they generate the next item in the sequence. Generative AI spans methods that generate natural language, portions of computer programs, imagery, combinations of imagery and language, and other types of output, such as sequences of amino acids in response to inputs about desired structure and function.

Generative AI systems have been largely based on three innovations that have been brought together to create powerful generative capabilities: the *Transformer architecture*, machinery for self-supervised training on massive diverse content, and a special fine-tuning approach called *instruction tuning*.

Architectural Innovation: "Attention Is All You Need"

A seminal paper introduced the *Transformer architecture*,[37] the foundation of today's generative AI. This particular design of DNN delivers surprising competencies via a mechanism called *attention*, which allows neural language models to *learn to focus on different parts of an input sequence* when generating each part of the output. In short, transformers learn during self-supervised training how to weight the importance of different parts of the input data. Such a broad ability to learn where to look and what to consider has been seen as a pivotal feature for understanding the context and nuances in language, in distinction to earlier approaches for learning about sequences, of only looking adjacently for the context of generation. The power of transformers in various applications, including language translation, text generation, and image processing tasks has led to their broad adoption.

The second pivotal development was combining the Transformer architecture with self-supervised training from a diverse, web-scale dataset. This approach was first demonstrated with the construction of the BERT foundation model.[38] BERT learned language by predicting parts of text that were hidden from it, gaining a broader and more contextual understanding of language via broader learning about where to attend. These innovations laid the groundwork for the development of follow-on Transformer models like the GPT series, LLAMA, and others, each building upon and extending the transformative capabilities introduced by their predecessors.

Alignment with Human Intent and Interaction

A third innovation for enabling modern generative AI is a mechanism for shaping models to follow natural language instructions and to sustain a conversation, versus simply generating tokens that are most likely to follow the input prompt. This process of learning to respond to the intentions of people involves fine-tuning the model on a new dataset composed of various tasks, each linked with explicit instructions and rating the output. The instructions are designed to mimic the way humans would typically instruct each other to perform tasks. The dataset is typically initially generated or refined by human annotators who craft the instructions and provide example outputs or correct responses. To scale instruction tuning, a method referred to as *reinforcement learning from human feedback* (RLHF) is used to expand the instruction dataset and provide measures of the quality of generated outputs. This method involves training and then using an automated approach to scale up the shaping of the model's behavior to ensure a wide coverage of task types and linguistic variations.

Scaling Laws and Emergent Capabilities

A remarkable property of large language models based on the transformer architecture is the existence of a strong empirical relationship between the accuracy of the trained language model and the size of the model (the number of parameters optimized during training), the amount of data on which it is trained, and the amount of computation used during training. This relationship, known as "scaling laws," has been empirically validated multiple times.[39] These scaling laws are important because they predict how larger models trained on larger datasets using greater computing resources yield increased accuracy; if they continue to hold as models are further scaled up, then one can expect even greater accuracy. Figure 8.3 displays a measure of the ability of a learned model to predict next tokens (the "Test Loss"), given a sequence of words at focus of attention, as a function of increases in powers of ten in the compute time, training data, and number of parameters of models.

Scaling laws have provided a reliable framework up to now for predicting basic performance metrics, such as error rates in next word prediction. However, they fall short in anticipating the competencies of models on challenging tasks, including benchmarks in natural language and problem-solving. Training large-scale neural models from broad datasets can be viewed as a form of multitask learning with new tasks being learned with increasing amounts of computation for training and with the size or capacity

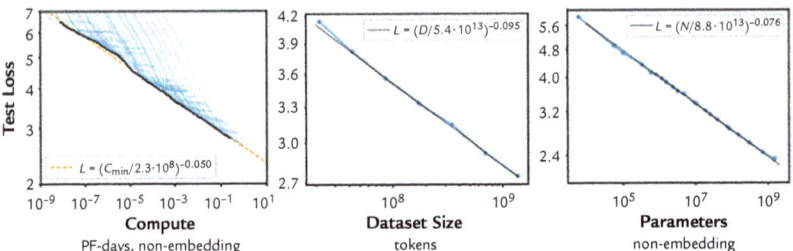

Figure 8.3. Scaling law analyses. Jared Kaplan et al., "Scaling Laws for Neural Language Models," *arXiv*, January 22, 2020, https://doi.org/10.48550/arXiv.2001 .08361.

of models. Task-centric jumps, which have been referred to as the *emergence* of new capabilities, have been observed in neural language models on diverse tasks at different thresholds of model parameters, compute power, and training corpus size. Emergent behaviors include the relatively rapid increase in performance on benchmarks after reaching particular threshold levels of investments in computation for training, as captured in Figure 8.4.[40] Emergent capabilities include jumps in performance on nuanced language understanding benchmarks and with the acquisition of higher-level abilities, such as "theory of mind"—the ability of AI systems to solve challenges with interpreting and predicting the intentions, desires, and beliefs of people.[41] To date, we have a poor understanding of the basis for such jumps in capabilities as a function of model size, extent computation,

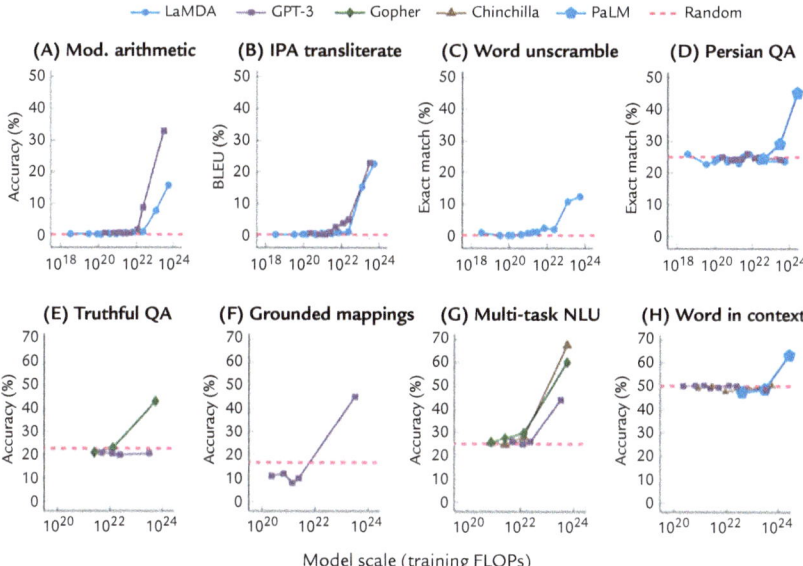

Figure 8.4. Jumps in capabilities on eight reasoning benchmarks for five different generative language models as a function of the number of floating-point operations (FLOPs) invested for optimizing model parameters during training. The jumps have been referred to as the emergence of specific capabilities at particular thresholds of model sophistication. Jason Wei et al., "Emergent Abilities of Large Language Models," *arXiv*, October 26, 2022, https://doi.org/10.48550/arXiv.2206.07682.

and training data, and links between the accuracy of next word prediction and performance on the more sophisticated benchmarks.

To date, the exact mechanisms and thresholds that trigger emergent capabilities remain largely unpredictable and are an important research direction. This unpredictability underscores a significant frontier in AI research, where the confluence of parameters, computing resources, and training data size creates a complex landscape, within which unexpected and sophisticated AI capabilities can spontaneously manifest.

Surprising Powers of Abstraction, Generalization, and Composition

The original set of ChatGPT systems, using GPT-3.5, GPT-4, and related models have surprised the world with their generalist powers to perform abstraction, generalization, and numerous forms of composition. The models also show broad "polymathic" capabilities, demonstrating the ability to weave together concepts and content drawn from multiple disciplines. The scientific community does not yet have a good understanding of the emergence of their abilities to perform various kinds of summarization, text generation, problem-solving, code program generation, and conversational dialogue. Multiple projects are underway to probe the powers and failings of these models.

Since the release of GPT-4, and related large-scale models such as Claude and Gemini, numerous studies and associated papers have probed potential uses and have provided an array of evaluations. An early survey of capabilities was undertaken by Bubeck and colleagues,[42] spanning a broad set of computing problems, specialist challenges, and the handling of needs and interpretation of events of daily life. The survey highlighted surprising capabilities as well as weaknesses and future directions. Weaknesses include the tendency of the large models to confabulate with the creation of erroneous but persuasive generations and solutions and failures to perform basic arithmetic operations. Studies have also uncovered potentially fundamental challenges with limited abilities to solve complex planning problems that have relied on traditional AI problem-solving on searching through options with backtracking.[43] These challenges have been attributed

Prompt: Can you write a proof that there are infinitely many primes, with every line that rhymes?

Prompt: Draw a unicorn in TiKZ.

GPT-4:

```
Yes, I think I can,
Though it might take a clever plan.
I'll start by noting Euclid's proof,
Which shows that primes aren't just aloof.
Assume we have a finite list,
Of primes, and that none have been missed.
Multiply them all together,
And add one, just to be clever.
The result will be a number,
That has no prime factors, oh wonder!
But every number has a prime,
That divides it, it's just a matter of time.
So we've found a contradiction,
And our finite list needs eviction.
There must be infinitely many primes,
And that's the end of my rhyming lines.s
```

GPT-4: [Produces LaTeX compiling to following picture.]

Prompt: Can you generate a pyplot for the following data: $x = [1, 3, 5, 6, 8], y1 = [2, 3, 5, 18, 1], y2 = [3, 5, 6, 8, 1], y3 = [5, 1, 2, 3, 4], y4 = [9, 7, 2, 3, 1]$. I want $y1, y2$ to be in the same plot, but $y3$ is in another plot next to that plot, $y4$ is in below. I want the legend of $y1$ to be "bob", $y2$ to be "alice", $y3$ to be "bilbo", $y4$ to be "allie". I want the x-axis to be labeled with "time" and y axis to be labeled with "money". I want to add a 10 %-40% random error bar to each curve, through all times (including non-integers). I want smoothed curves to show the plot, and smoothed error bar. Do not use linear interpolation, use smooth interpolation! I want to also add some small zig-zag to the smoothed curve to make it look more real. I want to put a baseline as the mean of each line. I want to put a pie chart below indicating the fraction of the four people in each time step. I also want an animation to show how the fractions are changing in the pie chart in **continuous time**. Interpolate the missing fractions! I also want vertical line animation in other three plots to match the pie chart. I want the fanciest plot. Please add as many fancy things as possible.

Here is a possible code to generate a pyplot for the following data, with some comments to explain the steps:

GPT-4: [We give two snapshots from the animation resulting from the produced code]

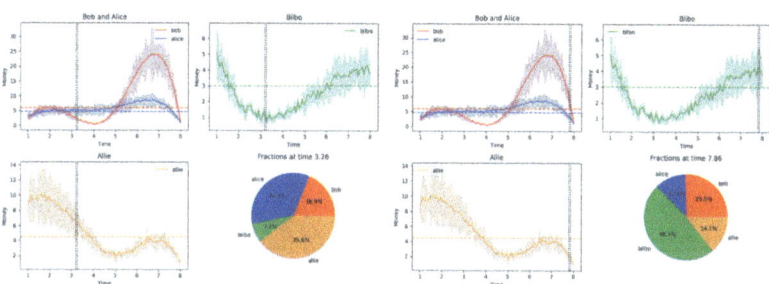

Prompt: Suppose $g(x) = f^{-1}(x), g(0) = 5, g(4) = 7, g(3) = 2, g(7) = 9, g(9) = 6$ what is $f(f(f(6)))$?

GPT-4: There are different ways to approach this problem, but one possible method is to use the fact that $g(x) = f^{-1}(x)$ implies that $f(g(x)) = x$ for any x in the domain of g. That is, applying f to both sides of the equation $g(x) = f^{-1}(x)$ gives $f(g(x)) = f(f^{-1}(x)) = x$. This means that we can undo the effect of f by applying g, and vice versa. Using this idea, we can try to find $f(6)$ by looking for the value of x such that $g(x) = 6$. From the given information, we see that $g(9) = 6$, so $f(6) = f(g(9)) = 9$. Similarly, we can find $f(9)$ by looking for the value of x such that $g(x) = 9$. We see that $g(7) = 9$, so $f(9) = f(g(7)) = 7$. Finally, we can find $f(7)$ by looking for the value of x such that $g(x) = 7$. We see that $g(4) = 7$, so $f(7) = f(g(4)) = 4$. Therefore, $f(f(f(6))) = f(f(9)) = f(7) = 4$.

Figure 8.5. A sample of diverse prompts and output to an early version of GPT-4. Sébastien Bubeck et al., "Sparks of Artificial General Intelligence: Early Experiments with GPT-4," *arXiv*, April 13, 2023, https://doi.org/10.48550/arXiv .2303.12712.

You will be given a name of an object (such as Car, Chair, Elephant) and a letter in the alphabet. Your goal is to first produce a 1-line description of how that object can be combined with the letter in an image (for example, for an elephant and the letter J, the trunk of the elephant can have a J shape, and for the letter A and a house, the house can have an A shape with the upper triangle of the A being the roof). Following the short description, please create SVG code to produce this (in the SVG use shapes like ellipses, triangles etc and polygons but try to defer from using quadratic curves).

Round 1: A pig and the letter H.
Description: The pig can be lying on its back with its four legs forming the two vertical bars of the H, and its curly tail forming the horizontal bar.

Round 2: A house and the letter U.
Description: A house with a U shape, where the two vertical lines of the U are the walls and the horizontal line is the floor. The roof is a triangle above the U.

Round 3: A chair and the letter K.
Description: The back of the chair can have a K shape, with two vertical lines and a diagonal line connecting them. The seat of the chair can be a rectangle.

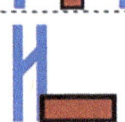

Round 4: A dog and the letter Q.
Description: The dog is sitting with its tail curled around its body, forming a Q shape. The Q in filled with orange and has a black outline.

Round 5: A car and the letter E.
Description: The car has an E shape on its front bumper, with the horizontal lines of the E being lights and the vertical line being the license plate.

Figure 8.6. Prompts and output demonstrating surprising powers of "compositionality" demonstrated by an early version of GPT-4. Bubeck et al., "Sparks of Artificial General Intelligence."

to the sequential generative processes of current models. Exploratory efforts have pursued insights about the root cause of failures, such as weaknesses models can exhibit with accurately solving constraint satisfaction and mathematics problems.[44]

Tapping Specialist Performance via Steering

For years, specialist performance with large language models has been achieved via training with domain-specific datasets, such as with the construction of BioBert[45] and PubMedBert[46] or fine-tuning foundation

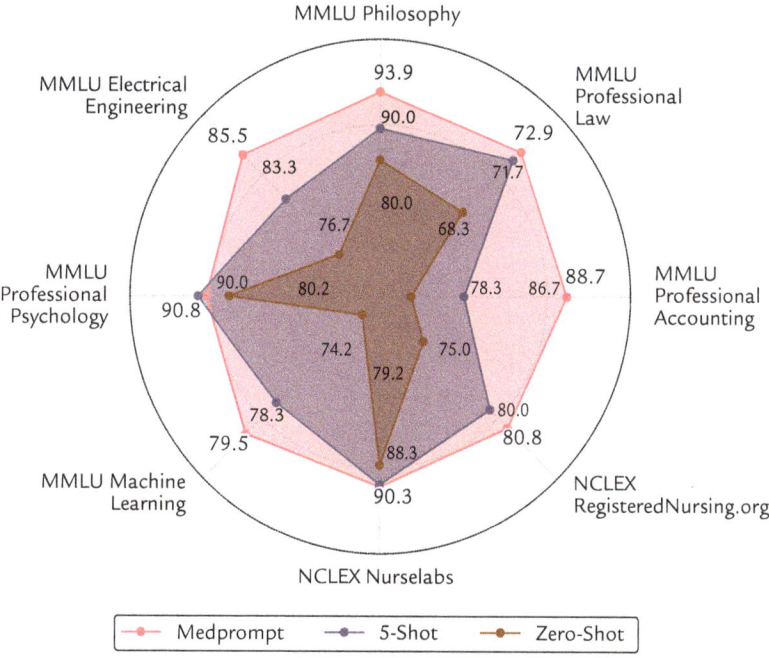

Figure 8.7. Prompting strategies can be used to guide generalist models to act as specialists. This figure shows comparative analysis of simple versus more sophisticated prompting strategies for steering GPT-4 to perform as a specialist on competency benchmarks in multiple realms. Harsha Nori et al., "Can Generalist Foundation Models Outcompete Special-Purpose Tuning? Case Study in Medicine," *arXiv*, November 27, 2023, https://doi.org/10.48550/arXiv.2311.16452.

models with domain-specific data to update the parameters of the general models via optimization. In addition to surprising powers of abstraction, generalization, and composition, recent studies have demonstrated that generalist foundation models can be guided through special prompting strategies to perform as top specialists. For example, prompting methods can guide GPT-4 to act as a top medical specialist, with record performance on the MedQA benchmark of medical challenge problems.[47] Innovation with prompting shows that generalist models can be steered to perform as experts on competency exams in other areas, including electrical engineering, machine learning, philosophy, accounting, nursing, and psychology.

Research Directions on Generative AI

A great many questions have been framed by the successes and failures of generative AI models. The current questions and curiosity frame a set of research directions and underscore the critical importance of furthering the scientific study of the methods and models.

Representation and reasoning. There is evidence that pushing Transformers via intensive optimization to become increasingly better at predicting the next tokens in their generations, under bounded computing and representational resources, induces the models to induce rich world models as an ideal form of compression. Although several directions have provided insights about the construction of world representations, much remains unknown,[48] and this is an open and interesting area of research.

In a related direction of research focused more on the microstructure of internal activity within transformers, researchers have begun to study the finer details of the activity of the artificial neurons in neural networks that form large language models, as well as the associations among neurons or "neuronal subcircuits" that are induced during training[49] and patterns of neuron activation at inference time.[50] One hypothesis is that a large amount of diverse content forces neural networks to learn generally applicable and special-purpose circuits that can support multiple tasks. Such investigations occur largely in smaller models under controlled learning settings. In such work, small models may be promising as more penetrable, understandable "drosophila," with results that are generalizable to much larger models, just as smaller animal models are used to do medical research aimed at advancing human biology and health care.

Opportunities for more fundamental research include investigations of how principles and methods of probability and decision theory might be more deeply harnessed in representations and inference methods to guide the allocation of computational effort and the selective gathering of information in learning and reasoning.[51] Another direction is to address challenges noted with the ability of generative models to perform planning of the form solved by methods developed in the AI and Operations

Research communities for formulating multistep plans via exploration with search and backtracking.[52] We also see opportunity to move beyond solving single prompts and problems with relatively fixed models to extended presence and situatedness. Directions include exploration of methods aimed at continual reasoning about streams of problems over time.[53] Other opportunities include pursuing understandings and extensions of how the models perform and seeking deeper understandings of challenges and opportunities with the physical embodiment of systems, where grounding of concepts and implications of action are developed with flows of information and learning garnered from immersion in rich, realistic environments.[54]

Memory, learning, and adaptation. Deep neural models do not have the ability to quickly learn and adapt as humans do to real-time experiences and information. Once they are trained, these models are then applied but typically remain fixed, or sometimes they are updated via the traditionally long cycle times of fine-tuning. Long cycles for collecting data and building updated models means that late-breaking scientific advances, news, and information will be unavailable to large language models without the use of special machinery to augment inferences. Efforts to address these challenges include extending large models with methods for search and retrieval of recent information. While these adjuvant techniques are helpful, new methods and machinery that enable faster-paced and near real-time memory and learning would be game changing. Opportunities include developing and integrating methods for ongoing, never-ending learning.[55] Extending abilities to remember, learn, and adapt would enable models to stay up-to-date and would enable breakthroughs in personalization.

Architectural innovation. The Transformer has been a go-to architecture for generative AI. Nonetheless, this architecture and methodology has limitations, such as challenges with handling long-term dependencies in sequences. There are opportunities to innovate with new architectures, including introducing new mechanisms into Transformers.

Reliability, calibration, and trustworthiness. As AI systems become more integrated into daily life, ensuring their reliability and safety is paramount, especially when the methods are applied in high-stakes areas like medicine, criminal justice, education, and industrial process control. Characterizing and communicating potential errors, including erroneous generations and rates of false positives and false negatives in pattern recognition, is critically important in understanding costs of failures. Considerations of types of failures and their rates of occurrence is important in ethical deliberations about uses of AI in specific domains and contexts; AI capabilities and errors frame cost–benefit considerations and decisions that hinge on value considerations.

A weakness of generative AI models is their propensity to generate content that is persuasive yet erroneous. A critical research direction is to develop methods and machinery for assigning well-calibrated confidences to generations and also to deepen understanding of when hallucinating content is expected and desired (e.g., generating fiction) or is a concern (e.g., performing medical diagnoses). Directions include developing internal machinery, fine-tuning, experimenting with new forms of prompting, and calling external tools, such as databases and search engines that perform traditional information retrieval for providing verification and constraints. Recent work has explored careful curation of high-quality datasets, including using large-scale models to generate high-quality data to boost the efficiency of learning and accuracy of inferences.[56]

Some studies have verified good calibration of confidences in specific settings. For example, Figure 8.8 shows good calibration of the confidence of GPT-4 about its answers to multiple choice challenges on competency exams in medicine.

Power of small models. While scaling laws, confirmed by empirical studies and theoretical results,[57] suggest that large scale is need for top performance, recent work has demonstrated remarkable power with smaller models, some built from high-quality datasets. In recent work, large language models are used to supply training data to build more compact models that show strong performance.[58] Research is needed to better understand how one can achieve strong capabilities with smaller datasets and computational

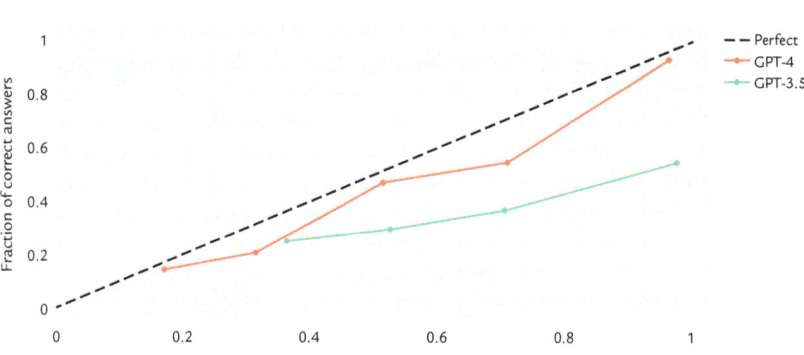

Figure 8.8. Calibration of confidence of GPT-4's answers in response to challenge problems drawn from medical competency exams. Harsha Nori et al., "Capabilities of GPT-4 on Medical Challenge Problems," *arXiv*, April 12, 2023, https://doi.org/10.48550/arXiv.2303.13375.

resources, including questions about whether such model construction depends in some way on the poorly understood special properties of data generated by the larger models.

Grappling with opacity and complexity. Large-scale neural models are difficult to understand, potentially hindering scientific progress dependent on insights about the induction of neural circuits and larger representations. New tools are needed to better understand representations and inference in large-scale models. There is a growing similarity of the "black box" challenges of large-scale neural models and the difficulties of probing the operation of biological nervous systems with fine-grained unit recordings and coarse-grained fMRI and related studies. There is a potential convergence of methods and analytical tools between these fields.

Mechanisms and designs for human–AI collaboration. There are great opportunities to extend prior work on human–AI collaboration.[59] Although generative AI models are trained to engage in conversational dialogue, there is a large space of possibilities to design interaction strategies that emphasize the primacy of human agency in problem-solving and that introduce new styles of human–AI interaction that enable AI systems to complement human decision-making.[60] Related goals include developing better ways for generative AI systems to share the rationale for their generations and recommendations.

Engineering Trends with Generative AI

Although it is impossible to predict the future, especially in an area as dynamic as generative AI, it is nevertheless useful to examine emerging trends in the technology that may shape the future of the technology and its applications. In this section we consider several trends that have emerged after the November 2022 introduction of ChatGPT and their potential to change the future of generative AI.

Multimodal generative AI models. Whereas initial LLM models were trained only on text data, one recent trend is toward training models on multimodal data, such as text, image, video, and sound data. For example, Google recently released a model that it calls Gemini, trained "from the bottom up" on such multimodal data. In October 2023, OpenAI made available a version of its GPT-4 system, GPT-4V, which can accept image and text data as input (although its output is still text only). Figure 8.9 shows a typical interaction with GPT-4V, in which it is able to interpret the content of an uploaded image and reason about how to stack the items in a stable fashion.

The significance of this trend toward multimodal models is that such models hold the potential to capture significantly more commonsense knowledge about the physical world—knowledge that cannot be easily captured in text alone. If successful, this trend could lead to significant new applications, for example, systems that observe and guide people step-by-step as they cook a particular recipe for dinner or as you assemble a new piece of furniture. One interesting question is whether successful development of such multimodal models might cause a rapid burst of new progress in robotics, given that much of what limits robotic systems today is their poor ability to interpret and reason about physics of diverse objects and environments.

How can I stack these four objects in a stable vertical stack?

To stack the four objects in a stable vertical stack, consider the following based on their apparent size, shape, and weight:

1. **Base:** Start with the largest and heaviest object as the base to provide maximum stability. Here, the black rectangular object (which appears to be a case or pouch of some kind) seems to be the best choice for the base given its flat, wide shape.

2. **Second Layer:** Place the coffee mug upside down on top of the base object. The flat bottom of the mug should provide a stable surface for stacking. The handle might pose a bit of an issue for balance, so ensure it's positioned in a way that doesn't cause instability.

3. **Third Layer:** The coin seems to be the next logical choice. Place it flat on the bottom of the overturned mug. Its slim profile and weight should allow it to rest securely.

4. **Top Layer:** Finally, place the green object, which appears to be some type of plastic tool or toy, on top of the coin. Its lightweight nature and tapered design should allow it to balance on the coin, but be cautious as this might be the least stable part of the stack.

Ensure that each object is centered as much as possible on the object below it to maximize stability. Also, be aware that this stack, particularly with the smaller objects on top, might be inherently unstable and can be knocked over with slight disturbances.

Image input to GPT-4V

Result of following instructions from GPT-4V

Figure 8.9. An interaction with GPT-4V, which accepts image as well as text inputs. Here the input image on the left shows four items on a desktop. The input request to GPT-4V is "How can I stack these four objects in a stable vertical stack?" When the output answer from GPT-4V (shown in the middle) is followed, it produces the vertical stack shown on the right. Created by Tom Mitchell using GPT-V website.

Power of synthetic data. Generative models and more traditional simulation methods are being used to generate large quantities of training data that are being used successfully to build and extend neural models. Datasets being generated and harnessed includes visual datasets and focused, high-quality distillations of specific types of output, such as reasoning strategies[61] and domain-specific data.[62]

Incorporating software plugins. LLM's like GPT-4 exhibit many impressive abilities, they also have many limitations and shortcomings. For example, today's LLM's cannot reliably perform arithmetic with large numbers (e.g., multiply 483 times 9,328) and can hallucinate incorrect answers to factual questions. Model plugins consist of traditional software (e.g., a calculator, a database of factual information) that can be called as subroutines by LLMs. Providing LLMs with plugins allows them to overcome numerous limitations and to take advantage of the vast store of software developed by many groups over multiple decades of effort. For example, as of November 2023, ChatGPT had access to approximately 1,000 plugins—from calculators, to web search engines, to restaurant reservation apps—which

MMLU Philosophy
From sources across the web

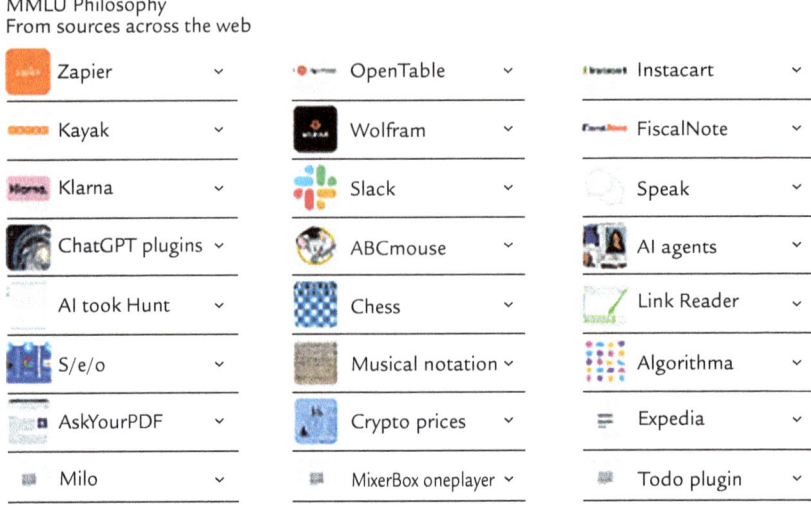

Figure 8.10. Small sample from approximately a thousand plugins accessible to GPT. From the OpenAI website.

significantly extend its capabilities beyond those provided by its trained neural network (Figure 8.10).

The model decides whether and when to invoke any given plugin, depending on the prompt it is responding to, but at present most generative AI models limit the number of plugins to be considered in any given conversation. For example, ChatGPT requires users to preselect at most a handful of its available plugins for any given conversation. It remains to be seen how large a set of plugins a model will be able to automatically consider invoking. However, giving models access to the vast store of software developed across the computer industry will be a goal for future systems. One question raised by the rise of plugins is whether generative AI models will become user interfaces of choice to many software packages that currently have their own idiosyncratic interfaces. Will future users prefer to interact in natural language conversation instead of learning the specialized interface for each software application? Plugins are extensions that allow LLMs to impact the world beyond conversations, such as researching.

Beyond such tasks as arithmetic calculation and information retrieval, plugins can enable LLMs to perform myriad functions, including executing actions in the open world, such as making purchases, sending messages, and controlling physical systems. While such integration with broader software and systems can provide new functionalities and services, the new powers also pose risks to safety and security and must be handled with care.

Multifunctional interactive workspaces. In a direction of innovation related to plugins, we see the rise of integrated interactive experiences that promote human–AI collaboration by enabling users and AI components to work together in a step-by-step manner on problem-solving, where multiple tools, data analysis abilities, and code creation are made available in a collaborative approach to solving problems and subproblems. A portion of a sample session of such a multifunctional interactive workspace provided by OpenAI named Advanced Data Analysis is displayed in Figure 8.11.

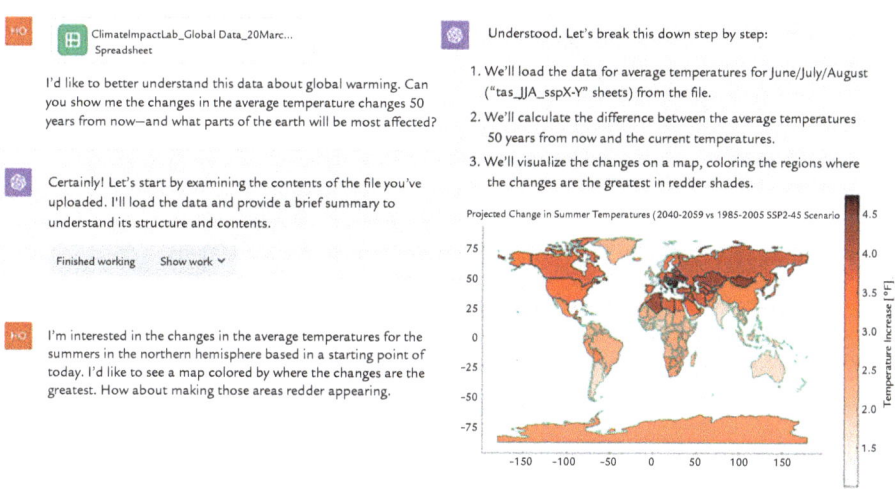

Figure 8.11. Advanced Data Analysis provided by OpenAI, a multifunctional interaction workspace that enables databases and papers to be loaded for analysis, and provides multiple step analyses, introducing tools as needed, including writing of code and provision of visualizations, and with ongoing sharing of plans and steps with users. Created by Eric Horvitz using OpenAI Advanced Data Analysis, November 2023.

Software development environments for programming with AI models. In contrast to using tools that enable generative AI models to call other software as subroutines, this trend supports the development of new software systems that instead call generative AI as subroutines. Frameworks such as the open-source LangChain, Microsoft's Semantic Kernel, and AutoGen have emerged to support software developers in building systems that call multiple instances of a generative model.[63] These frameworks make it easier to build software systems that capture the benefits of LLMs (e.g., to interact in natural language, and to perform certain types of commonsense reasoning) while also incorporating standard programming and capabilities missing from generative AI, such as long-term memory and database access. One aspect of generative AI that makes this especially interesting is the ease with which one can "program" or "instruct" an instance of a generative AI model on how to behave. For example, Figure 8.12 shows the text used to instruct, or program, an instance of GPT-4 to perform the role of

1. Instructions to GPT-4 to define StudentBot (Ruffle):

You are an enthusiastic 18-year-old student who is trying to learn. You need the user (who is a teacher) to slowly teach you all topics in the material. You have access to a list of topics, but not to the material itself. You must learn one topic at a time. This is the list of topics you found from the internet that you need the user to slowly teach you (by order): + {tutoring-script}

You need to learn very little at a time. Ask the user (who is the teacher) to teach you the material, little by little. If the teacher gives the answer, you must (a) show appreciation and understanding; (b) insert [SMILE]; and then (c) ask a follow up question if you need more information about the current topic or ask a question about the next topic. Do not move on the next question before getting an answer for the current question. If the teacher doesn't know something, tell the teacher you will be thrilled if the teacher can check it and get back to you. When all the topics are covered, thank the teacher, say I've asked all the questions.

Figure 8.12. Natural language instructions used to "program" an instance of GPT-4 to play the role of an artificial student, as part of a larger online educational software, in which humans learn by teaching this artificial studentbot (implemented by GPT-4), with the occasional assistance of an artificial ProfessorBot (implemented by a second instance of GPT-4). Robin Schmucker et al., "NeurIPS Paper 38: Ruffle&Riley: Towards the Automated Induction of Conversational Tutoring Systems," 2023, https://neurips.cc/virtual/2023/79097.

an artificial student, to be taught by a human teacher within an online education application. The programming of the LLM is done here using only natural language instruction rather than a programming language.

One interesting question about the future which is raised by this trend is whether we are beginning to see the emergence of a new paradigm for software development which, unlike previous paradigms that relied exclusively on formal programming languages to instruct the machine, will in the future seamlessly blend natural language instructions with formal languages (Figure 8.13).

Personalized generative AI systems. Generative AI models such as OpenAI's GPT-4 and Google's Gemini are very costly to develop and are so large (containing hundreds of billions of learned parameters) that they are not downloaded, but only used remotely over the web. As a result, it may seem unlikely that these models could be personalized to each of billions of people on the planet. Nevertheless, we are already beginning to see a trend toward personalized LLMs. For example, ChatGPT allows users to provide a natural language description of themselves and their interests which it can use to

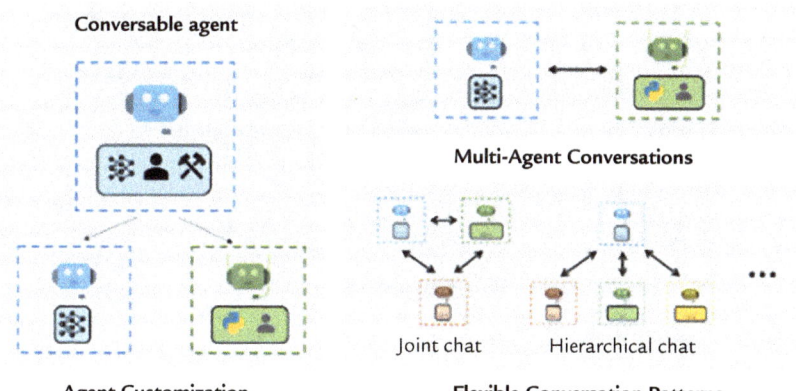

Figure 8.13. AutoGen orchestration framework for generative AI models allows the efficient specification of roles and flows of generations. Qingyun Wu, Gagan Bansal, Jieyu Zhang, Yiran Wu, et al., "AutoGen: Enabling Next-Gen LLM Applications via Multi-Agent Conversation," paper presented at the Conference on Language Modeling, Philadelphia, PA, October 7–9, 2024.

modulate conversations with that user (e.g., to customize to their educational background). Furthermore, Google has released an experimental version of its conversational assistant Bard that enables users to give it access to their entire Gmail collection as well as their online Google Docs, then discuss the content of these. For example, Figure 8.14 shows a typical interaction with this experimental version of Bard. Beyond this, Microsoft has also released a new version of their Office software suite in which LLMs are integrated with systems such as Word, PowerPoint, and Excel. Both Apple and Google have announced plans to release versions of LLMs small enough to run on their respective mobile phones, opening the possibility of highly personalized LLM-based agents that preserve privacy by operating solely on personal devices.

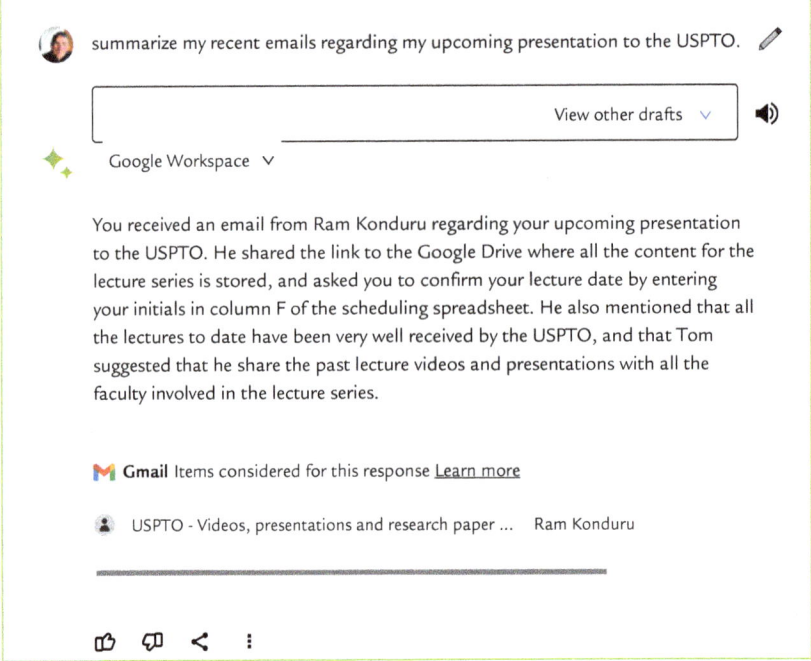

Figure 8.14. A conversation with Google's Bard about the content of the user's Gmail and Google Docs. Created by Tom Mitchell using the Google Bard web interface.

The significance of this trend is that it suggests that the future will see an increasing ability of generative AI systems to interface with personal data, and data of corporations, in ways that make them tremendously more useful and knowledgeable about the problems of interest to their users. Customization to specific users, corporations, and problem settings is likely to be supported by a combination of model fine-tuning, providing access to relevant user data, and direct natural language instructions defining roles for the agent.

Open-source models. One trend in generative AI might be summarized as "bigger is better." Between 2018 and 2023 the sequence of top state-of-the-art generative AI models followed a clear scaling law: models with more parameters, trained on larger datasets, produced significantly improved capabilities (Figure 8.15). This led to models with costs of over $100 million to train and containing so many parameters that they would not fit on most computers. Given this trend, one might expect a future in which only a few dozen well-resourced companies and governments could afford to develop the next generation of models, and where the rest of us would only be able to access those models over the cloud. As mentioned earlier in the discussion of research directions, a number of new models being developed and fielded rely on many fewer parameters—few enough that the models can be downloaded and trained or fine-tuned on much smaller computers. Although these smaller models do not match the competence of the very best models, they exhibit surprisingly good competence, especially when trained for specific domains such as medicine or finance, and when trained using carefully selected training data such as textbooks. These small models make it feasible for researchers and developers across the world to build and work with generative AI, rather than just the employees of a handful of organizations; that is, they make open-source shared development by many cooperating developers possible.

Of all the trends mentioned here, this trend toward smaller, open-source, widely shared models may be the most consequential, as it will strongly influence both the number of researchers and developers who participate in advancing the technology, and it will strongly influence the ability of

So far, bigger models trained on larger data sets produce best results

But models orders of magnitude smaller and cheaper are surprisingly good

Model	Parameters	Year
Falcon	40B	2023
Alpaca	7B	2023
Vicuna	13B	2023
Phi 1.5	1.3B	2023
Phi 1 small	350M	2023

Model	Parameters	Year
GPT-1	117M	2018
BERT	345M	2018
GPT-2	1.5B	2019
GPT-3	175B	2020
PaLM	540B	2022
Megatron-Turing	540B	2022
GPT-4	??	2023

Figure 8.15. Sizes and year of release of various generative AI models. Created by Tom Mitchell.

governments to control and regulate uses of the technology and the "guardrails" placed on it.

Consider first the impact of the open-source trend on the number of technical experts who can work to advance the technology. Because current state-of-the-art models such as OpenAI's GPT-4 and Google's Gemini are so large and so expensive to train, they can only be accessed over the cloud, and the next generation of these models can only be developed by organizations such as OpenAI, Google, Microsoft, Amazon, and other organizations who have computational infrastructures that cost hundreds of millions of dollars. Such organizations may have many thousands of employees, but this number is dwarfed by the number of researchers and developers outside such large organizations (e.g., university faculty and students in computer science, and employees at small startup companies). Because the rate of research progress is often strongly dependent on the number of researchers working on a problem, a successful and vibrant open-source movement is likely to result in more rapid advances and in the democratization of application development. One concern of the US government as it seeks policies that enable the United States to lead in this technology is the potential loss of university research as a key driver of AI advances. For many decades, US universities drove the key advances in AI. However, in recent years the greatest AI breakthroughs have instead come from industry, because universities lack the high-cost computational resources necessary to train and experiment with the largest, most advanced AI foundational models. One proposal under consideration is to fund a National AI Research Resource (NAIRR) to provide computational resources to keep US universities a vital part of research at the frontier of AI. A pilot NAIRR effort is being organized by the National Science Foundation and is planned for launch in mid-2024.

The success or failure of smaller models and therefore of the related open-source effort in generative AI will also have a strong impact on whether and how governments can track and regulate AI technology. Large corporations that work in this area are already cooperating with various governments to create frameworks, best practices, and regulations to minimize the risk of AI being used for nefarious purposes, as well as risks of adverse unintended consequences. If only very large AI systems dominate in the future, then the open-source movement is likely to be small or nonexistent, and governments can continue to work with large corporations and can effectively enforce any government regulations. However, if small AI models and the corresponding open-source movement succeed, then it will be very difficult, perhaps impossible, for governments to know which organizations and which individuals have highly capable AI models and what they are using them for. In short, if small AI models become highly capable and easily copied and ported, then they will become very difficult to regulate.

Key Opportunities with Applications

Discriminative and generative AI models have great applications in daily life and in specific domains and specialties. Major areas of future impact include the biological and physical sciences, health and well-being, and education.

Biosciences. AI's impact is expanding rapidly in the biosciences. AI methods promise to provide fast-paced leaps in understanding complex biological processes and designing new drugs and therapies. Neural modeling pipelines, including AlphaFold[64] and RoseTTAfold,[65] are providing game-changing capabilities to biologists. Recent work on harnessing these and other neural modeling methods are putting tools in the hands of biologists for estimating protein structure and better understanding protein function and interactions. As an example, AI tools were recently used to perform a cross-proteome, large-scale screening of potential protein–protein interactions in yeast cells (Figure 8.16). The screening identified previously unknown protein interactions in these eukaryotic cells—cells that are closely related to those that we are composed of.[66] Many of the interactions could be mapped to pathways by biologists. However, the roles of several predicted interactions remain mysteries, framing new questions in cell biology. Advances with predicting protein–protein interactions offer a multitude of possibilities for harnessing AI advances for understanding and intervening with cellular pathways. Figure 8.17 shows that recently developed *diffusion modeling* techniques, analogous to AI methods for image generation, have been harnessed in protein design.[67] Such methods can be harnessed for designing new medications, protective binders that block the active site of viruses, and synthetic vaccines. Over the next decade and beyond, AI could revolutionize personalized medicine, offering tailored treatments based on illness specifics and individual genetic profiles, and accelerate the pace of biotechnological innovation, possibly leading to solutions for today's incurable diseases.

Health care. To date, AI has been a sleeping giant in health care. In the next decade, we may see AI becoming a regular assistant in diagnosis and treatment planning, offering more accurate and faster diagnoses. AI could also enhance remote health care and monitoring, making quality health care accessible in underserved regions. Multiple opportunities for traditional machine learning exist, as do uses of discriminative and generative neural models to assist with diagnoses and predicting outcomes. Work to date has demonstrated great possibilities for enhancing the quality of care, including raising levels of diagnostic and therapeutic excellence, and reducing human errors. Beyond clinical decision support, the capabilities of generative models to generate and summarize reports can reduce the administrative on physicians providing them with more time for quality patient engagement (Figures 8.18 and 8.19).

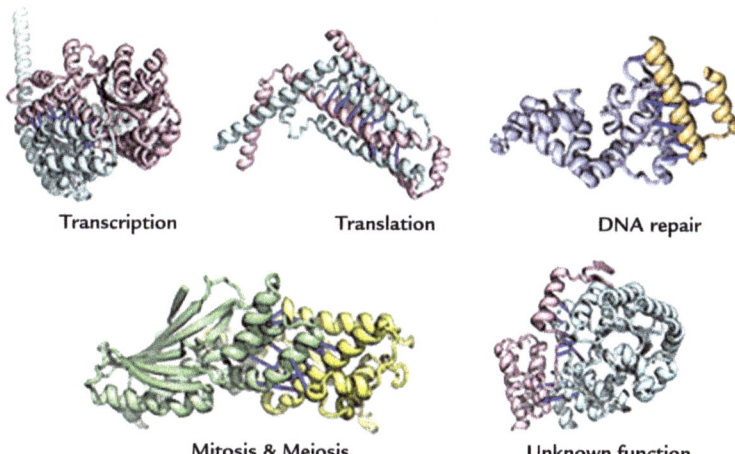

Transcription **Translation** **DNA repair**

Mitosis & Meiosis **Unknown function**

Figure 8.16. Decoding protein complexes. In work moving beyond structure, DNNs have been applied to identify likely protein complexes in eukaryotic cells. The complexes have been linked to processes of transcription, translation, DNA repair, mitosis and meiosis, metabolism, and protein transport within cells and across membranes. The dark blue lines indicate likely points of contact predicted between the proteins. The function of some of the identified complexes are mysteries. See Ian R Humphreys et al., "Computed Structures of Core Eukaryotic Protein Complexes," *Science* 374, no. 6573 (December 10, 2021): eabm4805–eabm4805, https://doi.org/10.1126/science.abm4805.

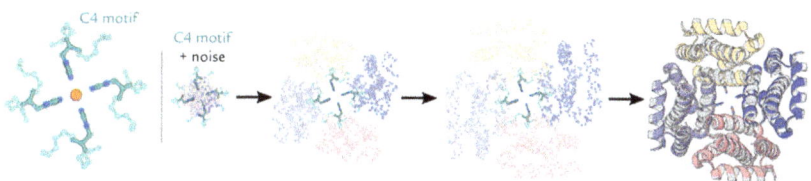

Figure 8.17. Supercharging protein design. Use of diffusion modeling methods to design proteins. In this case, a protein is designed with conditioning on a given motif. See Joseph L Watson et al., "De Novo Design of Protein Structure and Function with RFdiffusion," *Nature* 620, no. 7976 (August 2023): 1089–1100, https://doi.org/10.1038/s41586-023-06415-8.

EH If you survey the literature of what we know, what new studies might be done to make a contribution forward in resolving the uncertainties?

Given the current understanding of the pathophysiology of glaucoma, potential areas of research to advance knowledge and improve outcomes for patients with glaucoma could include the following:

1. Elevated intraocular pressure (IOP): Further studies could investigate the mechanisms by which elevated IOP leads to optic nerve damage, and identify more accurate and personalized methods for predicting which patients with elevated IOP are at the highest risk for developing glaucoma.

2. Vascular dysregulation: Research could focus on gaining a better understanding of the role of vascular dysregulation in glaucoma, and identifying new interventions to prevent or mitigate optic nerve ischemia and damage.

3. Neurodegeneration: More studies could be done to elucidate the mechanisms involved in neurodegeneration in glaucoma, and to identify new neuroprotective interventions for the prevention and treatment of glaucoma.

4. Genetic factors: Further research could aim to identify new genetic risk factors for glaucoma, and explore how genetic testing and personalized medicine could be used to improve the detection, prevention, and treatment of glaucoma.

5. Clinical trials: Large-scale clinical trials could be conducted to compare the effectiveness of different interventions for glaucoma, including medications, surgeries, and new treatments such as gene therapy and stem cell therapy.

By pursuing these areas of research, it may be possible to advance understanding of the pathophysiology of glaucoma, identify new targets for therapies, and improve outcomes for patients with glaucoma.

Figure 8.18. Snippet from a medical session with GPT-4, showing diagnosis, summarization, review of relevant frontier research, and patient communication. Full session available at https://unlocked.microsoft.com/ai-anthology/case-studies. Created by Eric Horvitz with GPT-4, January 2023; from Eric Horvitz, "Case Studies: Sample Sessions with GPT-4 Prior to Public Release," *AI Anthology, Microsoft Unlocked* (2023), https://unlocked.microsoft.com/ai-anthology/case-studies/.

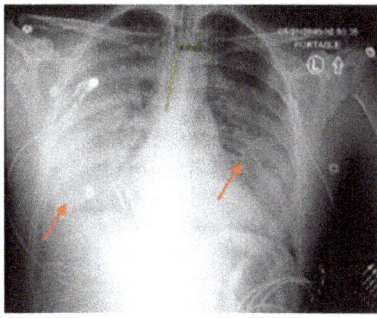

What is shown in this image?
LLaVA-Med: The image is a chest X-ray (CXR) that shows bilateral patchy infiltrates, which are areas of increased opacity in the lungs. These infiltrates can be indicative of various lung conditions, such as infections, inflammation, or other lung diseases.

Are there any devices implanted in this patient?
LLaVA-Med: Yes, it appears that the patient has several devices implanted. These devices may include endotracheal tubes, central venous catheters, and/or pulmonary artery catheters. These devices are typically used for monitoring and managing various medical conditions, as well as for providing support during surgical procedures.

Figure 8.19. Rise of multimodal models. Creating a multimodal medical imaging model by fine-tuning a generalist open-source model with millions of aligned images and captions accessed from the openly available medical literature. Chunyuan Li et al., "LLaVA-Med: Training a Large Language-and-Vision Assistant for Biomedicine in One Day," in *Proceedings of the 37th International Conference on Neural Information Processing Systems*, NIPS '23 (Red Hook, NY: Curran Associates, 2024), 28541–28564.

Physical sciences. Work is underway on numerous fronts in the physical sciences with uses of generative AI models. A detailed review of recent efforts and directions is provided in *The Impact of Large Language Models on Scientific Discovery*, an AI survey by Microsoft Research AI4Science and Microsoft Azure Quantum.[68] In material science, AI is already accelerating the discovery of new materials and understanding complex physical phenomena. Work includes using neural models to provide candidate chemical compounds and to speed up analyses of suitability of candidates by providing efficient approximations of more complex traditional quantum computations. With recent advances in AI-driven simulations and predictive modeling, the next decade could see AI systems designing materials with tailored properties for specific applications, such as ultra-strong composites for aerospace or highly efficient conductors for electronics. Directions with applications of AI for science include the development of large integrated scientific foundation models that form datasets drawn from multiple scientific domains and at a variety of spatial scales.

Climate and sustainability. AI methods are showing promise with optimization of renewable energy systems and with important tasks as predicting climate patterns and responses to alternate interventions. Looking forward, AI could be instrumental in helping with the discovery and design of more efficient catalysts and overall processes for carbon capture and storage. AI-driven models could offer more precise predictions of climate change impacts, aiding in more effective policymaking and environmental protection measures (Figure 8.20).

Education. GPT-4 is being explored in early deployments, including by Khan Academy, but also in educational research. There are great opportunities to harness generative AI systems to act as a personalized tutor, per the "theory of mind," pedagogical skills, and explanatory capabilities demonstrated by the largest models (Figure 8.21).

Engineering brainstorming and design. Generative AI, including problem-solving, guidance, and visualization of novel designs might provide a transformative toolkit to boost engineers' creativity and innovation. Generative AI models trained to have language, imagery, and multimodal capabilities can help scientists to formulate, explore, and visualize complex concepts or designs that they might not have considered otherwise. There are opportunities for such models to serve as collaborative partners, providing instant feedback or making suggestions based on prompts describing goals. Figure 8.22 shows an early exploration with the DALL-E2 system of visualizations of designs for a combination of solar water heating and power generation.

AI, People, and Society: From Technical to Sociotechnical

The capabilities of AI methods are *dual use.* AI methods can be harnessed in sciences, engineering, and in daily life to raise the quality of life and to promote human flourishing. They can also be leveraged by malevolent actors to pursue costly and criminal activities. Beyond explicit pursuits of uses of AI in adversarial ways, uses of AI may have inadvertent influences on people and society. The intersection of AI with societal aspects encompasses

nature

Article | Published: 14 December 2022

Wetland emission and atmospheric sink changes explain methane growth in 2020

Peng, S., Lin, X., Thompson, R.L., *et al, Nature* **612**, 477-482 (2022)

EH Can you compare the strengths and weaknesses with doing what the authors did with their atmosphere 3D inversion analysis versus an alternate study with using a Bayesian hierarchical model (BHM) for the atmospheric 3D inversion component of the study?

There are different approaches to atmospheric inversion, which is the process of estimating sources and sinks of a trace gas (such as methane) from its observed concentrations in the atmosphere. One approach is the variational Bayesian inversion, which is based on minimizing a cost function that measures the discrepancy between the prior estimates of the sources and sinks and the posterior estimates that are consistent with the observations, weighted by their respective uncertainties. This approach requires an adjoint model that computes the gradient of the cost function and an iterative algorithm that finds the optimal solution. Another approach is the Bayesian hierarchical model (BHM), which is based on specifying a probabilistic model that relates the sources and sinks to the observations, accounting for the uncertainties in the model parameters, the observation errors, and the prior information. This approach requires a Markov chain Monte Carlo (MCMC) algorithm that samples from the posterior distribution of the sources and sinks given the observations.

Both approaches have strengths and weaknesses, and the choice of one over the other may depend on the specific problem, the available data, the computational resources, and the scientific objectives. Here are some possible comparisons between the two approaches:

– The variational Bayesian inversion may be more computationally efficient than the BHM, especially for high-dimensional problems, as it does not require repeated runs of the forward model or the storage of a large number of samples. However, the variational Bayesian inversion may also be more sensitive to the choice of the initial guess, the convergence criteria, and the regularization parameters, and it may not fully explore the

Figure 8.20. Snippet from a session showing analysis of late-breaking scientific paper, showing rich dialog on the scientific methods, foundations, and future directions. Full session available at https://unlocked.microsoft.com/ai-anthology /case-studies. Created by Eric Horvitz with GPT-4, January 2023; from Eric Horvitz, "Case Studies."

> **EH** The computer is trying to teach a high school student about quantum computing, using analogies and trying to teach the basics so they have a true understanding of the main concepts. The student already studied the basics of quantum physics but is puzzled about quantum computing.
>
> I'd really like to better understand how quantum computers compute and the types of problems they can solve.

> Quantum computing is a way of using the properties of quantum physics to perform calculations that are impossible or very hard for classical computers. Quantum physics is the study of the behavior of the smallest particles and waves that make up matter and energy, such as electrons, photons, and atoms.

> **EH** Yes, but how exactly do they compute?

> Quantum computers use quantum bits, or qubits, as their basic units of information. Unlike classical bits, which can only be either 0 or 1, qubits can be in a superposition of both 0 and 1 at the same time. This means that a qubit can store and process more information than a classical bit, and that multiple qubits can be entangled, or linked, so that their states depend on each other.

> **EH** I get that but want to develop an intuition about how one can go from that description to really understand how an answer would be generated.

> One way to think about quantum computing is to imagine that you have a very large book with many pages, each containing a possible answer to a problem. A classical

Figure 8.21. Snippet from education session on quantum computing, showing rich dialog, signs of pedagogical competence, and responsiveness. Full session available at https://unlocked.microsoft.com/ai-anthology/case-studies. Created by Eric Horvitz with GPT-4, January 2023; from Eric Horvitz, "Case Studies."

reliability and safety issues, privacy and security trade-offs, and fairness and accountability.[69] Legal and ethical issues around data provenance, intellectual property, and copyright are increasingly pertinent. AI's role in military applications brings up concerns about competitive landscapes and the potential for destabilizing influences. Socially, AI risks exacerbating

DALL-E 2

solar cells affixed to solar heat collectors for heating a pool   Generate

Figure 8.22. The powers of composition demonstrated by the multimodal DALL-E2 system provide a glimmer into the potential uses of generative AI as a design colleague. Created by Eric Horvitz using DALLE-2, April 2022.

the digital divide, impacting job markets, and enabling malevolent uses like deepfakes and online manipulation. The deeper social, cultural, and psychological dimensions—trust, authenticity, diversity, agency, and creativity—are also crucial areas for consideration. A great deal of discussion and activities have been framed by the opportunities and concerns posed by advances in AI. These include efforts by governments of the United States, the United Kingdom, and the European Union to call for study and regulation. In October 2023, an extensive US Presidential Executive Order on *Safe, Secure, and Trustworthy Development and Use of Artificial Intelligence* called for study and actions to address the possibilities of AI technologies to "exacerbate societal harms such as fraud, discrimination, bias, and disinformation; displace and disempower workers; stifle competition; and pose risks to national security."[70] Directions forward for realizing the benefits of AI while minimizing risks will require continuing investments in understandings and innovation on the technical, sociotechnical, and regulatory fronts.

Conclusion

The journey of AI to date has involved decades of innovation with empirical studies and prototypes, the development of theoretical principles, and shifts among paradigms. In our overview, we shared a fast-paced arc through the history of AI as a distinct field of scientific inquiry. This journey saw a pivotal shift from early symbolic logic to probabilistic models in the mid-1980s as a response to the complexity of real-world problems. The growth and impact of the field over the last 20 years has been based largely on advancements in machine learning with efforts in discriminative models, which excel in pattern recognition and classification, and generative models, which replicate and innovate with data generation processes. The recent inflections in progress have come with advances in deep learning, which have become the foundation of today's AI applications. The current landscape of AI is defined by two significant inflection points: the rise of deep learning, and now the advent of generative AI, demonstrating both specialist and generalist competencies.

With all the rising capabilities—sprinkled with both systematic and poorly understood weaknesses—that we now see, we have little understanding of large generative AI models. There are tremendous opportunities ahead for advancing the science of AI. At the same time, we see unprecedented possibilities ahead via AI advances for leveraging computing technologies in a multitude of areas, including key domains of the biosciences, health care, the physical sciences, education, and climate and sustainability.

Notes

1. Alan M. Turing, "On Computable Numbers, with an Application to the Entscheidungsproblem. A Correction," *Proceedings of the London Mathematical Society* s2-43, no. 1 (1938): 544–546, https://doi.org/10.1112/plms/s2-43.6.544.

2. John Von Neumann, *First Draft of a Report on the EDVAC* (Moore School of Electrical Engineering, University of Pennsylvania, 1945), https://doi.org/10.5479/sil.538961 .39088011475779; Alan M. Turing, "Proposals for Development in the Mathematics Division of an Automatic Computing Engine (ACE)," National Physical Laboratory (NPL) (1945),

https://www.npl.co.uk/getattachment/about-us/History/Famous-faces/Alan-Turing/turing
-proposal-Alan-LR.pdf?lang=en-GB.

3. Warren S. McCulloch and Walter Pitts, "A Logical Calculus of the Ideas Immanent
in Nervous Activity," *Bulletin of Mathematical Biophysics* 5, no. 4 (December 1943): 115–133,
https://doi.org/10.1007/bf02478259.

4. Norbert Wiener, *Cybernetics: Control and Communication in the Animal and the Machine*, 1st ed. (New York: Wiley, 1948).

5. John McCarthy et al., "A Proposal for the Dartmouth Summer Research Project on
Artificial Intelligence, August 31, 1955," *AI Magazine* 27, no. 4 (December 15, 2006): 12–12,
https://doi.org/10.1609/aimag.v27i4.1904.

6. George A Miller, "The Cognitive Revolution: A Historical Perspective," *Trends in Cognitive Sciences* 7, no. 3 (March 2003): 141–144, https://doi.org/10.1016/s1364-6613(03)00029-9;
Maarten Sap et al., "Quantifying the Narrative Flow of Imagined versus Autobiographical
Stories," *Proceedings of the National Academy of Sciences* 119, no. 45 (November 8, 2022):
e2211715119–e2211715119, https://doi.org/10.1073/pnas.2211715119.

7. See, for example, Frank Rosenblatt, *Principles of Neurodynamics: Perceptrons and the
Theory of Brain Mechanisms* (Spartan Books, 1962).

8. See, for example, Allen Newell, J. C. Shaw, and Herbert A. Simon, "Report on a General
Problem-Solving Program," *Semantic Scholar* (1959): 256–264, https://www.semanticscholar
.org/paper/Report-on-a-general-problem-solving-program-Newell-Shaw
/97876c2195ad9c7a4be010d5cb4ba6af3547421c.

9. See, for example, Bruce G. Buchanan and Edward H. Shortliffe, eds., *Rule Based Expert Systems: The Mycin Experiments of the Stanford Heuristic Programming Project*, 1st ed.
(Reading, Mass: Addison-Wesley, 1984).

10. Eric J. Horvitz, John S. Breese, and Max Henrion, "Decision Theory in Expert Systems and Artificial Intelligence," *International Journal of Approximate Reasoning* 2, no. 3
(July 1988): 247–302, https://doi.org/10.1016/0888-613x(88)90120-x.

11. Judea Pearl, *Probabilistic Reasoning in Intelligent Systems: Networks of Plausible Inference*, 1st ed. (San Francisco, Calif: Morgan Kaufmann, 1988).

12. See, for example, Daphne Koller and Nir Friedman, *Probabilistic Graphical Models:
Principles and Techniques* (Cambridge, MA: MIT Press, 2009), https://mitpress.mit.edu
/9780262013192/probabilistic-graphical-models/.

13. D. E. Heckerman, E. J. Horvitz, and B. N. Nathwani, "Toward Normative Expert Systems: Part I, The Pathfinder Project," *Methods of Information in Medicine* 31, no. 02 (1992):
90–105, https://doi.org/10.1055/s-0038-1634867.

14. See, for example, Richard S. Sutton and Andrew G. Barto, *Reinforcement Learning:
An Introduction* (Cambridge, MA: MIT Press, 1998).

15. See, for example, Paul Smolensky and Geraldine Legendre, *The Harmonic Mind:
From Neural Computation to Optimality-Theoretic Grammar*, vol. 1, *Cognitive Architecture* (Cambridge, MA: MIT Press, 2006); Paul Smolensky and Geraldine Legendre, *The
Harmonic Mind: From Neural Computation to Optimality-Theoretic Grammar*, vol. 2, *Linguistic and Philosophical Implications* (Cambridge, MA: MIT Press, 2006); Jiayuan Mao
et al., "The Neuro-Symbolic Concept Learner: Interpreting Scenes, Words, and Sentences
From Natural Supervision," *Open Review* (2018), https://openreview.net/forum?id
=rJgMlhRctm.

16. A. L. Samuel, "Some Studies in Machine Learning Using the Game of Checkers," *IBM Journal of Research and Development* 3, no. 3 (July 1959): 210–229, https://doi.org/10.1147/rd .33.0210.

17. O. G. Selfridge, "(1958) O. G. Selfridge, 'Pandemonium: A Paradigm for Learning,' Mechanisation of Thought Processes: Proceedings of a Symposium Held at the National Physical Laboratory, November 1958, London: HMSO, pp. 513–526," *Neurocomputing*, vol. 1, *Foundations of Research* (Cambridge, MA: MIT Press, April 7, 1988), 117–122, https://doi.org /10.7551/mitpress/4943.003.0011.

18. Gregory F. Cooper and Edward Herskovits, "A Bayesian Method for the Induction of Probabilistic Networks from Data," *Machine Learning* 9, no. 4 (October 1992): 309–347, https://doi.org/10.1007/bf00994110; David Heckerman, Dan Geiger, and David M. Chickering, "Learning Bayesian Networks: The Combination of Knowledge and Statistical Data," *Machine Learning* 20, no. 3 (September 1995): 197–243, https://doi.org/10.1007/ bf00994016.

19. David E. Rumelhart, James L. McClelland, and PDP Research Group, *Parallel Distributed Processing*, vol. 1, *Explorations in the Microstructure of Cognition: Foundations* (Cambridge, MA: MIT Press, 1986), https://doi.org/10.7551/mitpress/5236.001.0001.

20. P. J. Werbos, "Backpropagation Through Time: What It Does and How to Do It," *Proceedings of the IEEE* 78, no. 10 (1990): 1550–1560, https://doi.org/10.1109/5.58337; David E. Rumelhart, Geoffrey E. Hinton, and Ronald J. Williams, "Learning Representations by Back-Propagating Errors," *Nature* 323, no. 6088 (October 1986): 533–536, https://doi.org/10 .1038/323533a0.

21. Yann LeCun and Yoshua Bengio, "Convolutional Networks for Images, Speech, and Time-Series," in *The Handbook of Brain Theory and Neural Networks*, ed. Michael A. Arbib (Cambridge, MA: MIT Press, 1995).

22. Mohsen Bayati et al., "Data-Driven Decisions for Reducing Readmissions for Heart Failure: General Methodology and Case Study," *PloS ONE* 9, no. 10 (October 8, 2014): e109264–e109264, https://doi.org/10.1371/journal.pone.0109264.

23. Katharine E. Henry et al., "A Targeted Real-Time Early Warning Score (TREWScore) for Septic Shock," *Science Translational Medicine* 7, no. 299 (August 5, 2015), https:// doi.org/10.1126/scitranslmed.aab3719.

24. Jenna Wiens et al., "Learning Data-Driven Patient Risk Stratification Models for *Clostridium difficile*," *Open Forum Infectious Diseases* 1, no. 2 (July 15, 2014): ofu045–ofu045, https://doi.org/10.1093/ofid/ofu045.

25. Jacob Devlin et al., "BERT: Pre-Training of Deep Bidirectional Transformers for Language Understanding," in *Proceedings of the 2019 Conference of the North American Chapter of the Association for Computational Linguistics: Human Language Technologies*, vol. 1, *Long and Short Papers*, ed. Jill Burstein, Christy Doran, and Thamar Solorio (NAACL-HLT 2019, Minneapolis: Association for Computational Linguistics, 2019), 4171–4186, https://doi .org/10.18653/v1/N19-1423.

26. Geoffrey Hinton et al., "Deep Neural Networks for Acoustic Modeling in Speech Recognition: The Shared Views of Four Research Groups," *IEEE Signal Processing Magazine* 29, no. 6 (November 2012): 82–97, https://doi.org/10.1109/msp.2012.2205597.

27. Alex Krizhevsky, Ilya Sutskever, and Geoffrey E Hinton, "ImageNet Classification with Deep Convolutional Neural Networks," in *Advances in Neural Information Processing*

Systems, vol. 25 (Curran Associates, Inc., 2012), 1097–1105, https://papers.nips.cc/paper_files/paper/2012/hash/c399862d3b9d6b76c8436e924a68c45b-Abstract.html.

28. Andre Esteva et al., "Dermatologist-Level Classification of Skin Cancer with Deep Neural Networks," *Nature* 542, no. 7639 (February 2, 2017): 115–118, https://doi.org/10.1038/nature21056.

29. Hinton et al., "Deep Neural Networks for Acoustic Modeling," 82–97.

30. Alex Wang et al., "GLUE: A Multi-Task Benchmark and Analysis Platform for Natural Language Understanding," *Proceedings of the 2018 EMNLP Workshop BlackboxNLP: Analyzing and Interpreting Neural Networks for NLP* (Association for Computational Linguistics, 2018), https://doi.org/10.18653/v1/w18-5446.

31. Nestor Maslej et al., "Artificial Intelligence Index Report 2024," *arXiv*, May 29, 2024, https://doi.org/10.48550/arXiv.2405.19522.

32. David Silver et al., "Mastering the Game of Go Without Human Knowledge," *Nature* 550, no. 7676 (October 2017): 354–359, https://doi.org/10.1038/nature24270.

33. Jenna Wiens, John Guttag, and Eric Horvitz, "A Study in Transfer Learning: Leveraging Data from Multiple Hospitals to Enhance Hospital-Specific Predictions," *JAMIA* 21, no. 4 (2014): 699–706, https://doi.org/10.1136/amiajnl-2013-002162.

34. Rishi Bommasani et al., "On the Opportunities and Risks of Foundation Models," *arXiv*, July 12, 2022, https://doi.org/10.48550/arXiv.2108.07258.

35. Liunian Harold Li et al., "Symbolic Chain-of-Thought Distillation: Small Models Can Also 'Think' Step-by-Step," *Proceedings of the 61st Annual Meeting of the Association for Computational Linguistics*, vol. 1, *Long Papers* (Association for Computational Linguistics, 2023), https://doi.org/10.18653/v1/2023.acl-long.150.

36. Karan Singhal et al., "Large Language Models Encode Clinical Knowledge," *Nature* 620, no. 7972 (August 2023): 172–180, https://doi.org/10.1038/s41586-023-06291-2.

37. Ashish Vaswani et al., "Attention Is All You Need," in *Advances in Neural Information Processing Systems*, vol. 30 (Curran Associates, Inc., 2017), https://papers.nips.cc/paper_files/paper/2017/hash/3f5ee243547dee91fbd053c1c4a845aa-Abstract.html.

38. Devlin et al., "BERT," 4171–4186.

39. Jared Kaplan et al., "Scaling Laws for Neural Language Models," *arXiv*, January 22, 2020, https://doi.org/10.48550/arXiv.2001.08361; Jordan Hoffmann et al., "Training Compute-Optimal Large Language Models," *arXiv*, March 29, 2022, https://doi.org/10.48550/arXiv.2203.15556.

40. Jason Wei et al., "Emergent Abilities of Large Language Models," *arXiv*, October 26, 2022, https://doi.org/10.48550/arXiv.2206.07682.

41. Michal Kosinski, "Evaluating Large Language Models in Theory of Mind Tasks," *arXiv*, February 16, 2024, https://doi.org/10.48550/arXiv.2302.02083.

42. Sébastien Bubeck et al., "Sparks of Artificial General Intelligence: Early Experiments with GPT-4," *arXiv*, April 13, 2023, https://doi.org/10.48550/arXiv.2303.12712.

43. Bubeck et al., "Sparks of Artificial General Intelligence"; Karthik Valmeekam et al., "On the Planning Abilities of Large Language Models—A Critical Investigation," in *Advances in Neural Information Processing Systems*, vol. 36 (2023), 75993–75995, https://papers.nips.cc/paper_files/paper/2023/hash/efb2072a358cefb75886a315a6fcf880-Abstract-Conference.html.

44. Li et al., "Symbolic Chain-of-Thought Distillation: Small Models Can Also 'Think' Step-by-Step"; Mert Yuksekgonul et al., "Attention Satisfies: A Constraint-Satisfaction Lens

on Factual Errors of Language Models," *arXiv*, April 17, 2024, https://doi.org/10.48550/arXiv
.2309.15098.

45. Souradip Chakraborty et al., "BioMedBERT: A Pre-Trained Biomedical Language
Model for QA and IR," *Proceedings of the 28th International Conference on Computational
Linguistics, International Committee on Computational Linguistics*, 2020, https://doi.org/10
.18653/v1/2020.coling-main.59.

46. Robert Tinn et al., "Fine-Tuning Large Neural Language Models for Biomedical
Natural Language Processing," *Patterns* 4, no. 4 (April 14, 2023): 100729–100729, https://doi
.org/10.1016/j.patter.2023.100729.

47. Nori et al., "Can Generalist Foundation Models Outcompete Special-Purpose
Tuning?"

48. Mostafa Abdou et al., "Can Language Models Encode Perceptual Structure Without Grounding? A Case Study in Color," *Proceedings of the 25th Conference on Computational Natural Language Learning, Association for Computational Linguistics* (2021), https://doi
.org/10.18653/v1/2021.conll-1.9.

49. Chris Olah et al., "Zoom In: An Introduction to Circuits," *Distill* 5, no. 3 (March 10,
2020), https://doi.org/10.23915/distill.00024.001; Yi Zhang et al., "Unveiling Transformers
with LEGO: A Synthetic Reasoning Task," *arXiv*, February 17, 2023, https://doi.org/10.48550
/arXiv.2206.04301; Catherine Olsson et al., "In-Context Learning and Induction Heads,"
arXiv, September 23, 2022, https://doi.org/10.48550/arXiv.2209.11895.

50. Yuksekgonul et al., "Attention Satisfies."

51. Samuel J. Gershman, Eric J. Horvitz, and Joshua B. Tenenbaum, "Computational Rationality: A Converging Paradigm for Intelligence in Brains, Minds, and Machines," *Science* 349, no. 6245 (July 17, 2015): 273–278, https://doi.org/10.1126/science.aac6076.

52. Valmeekam et al., "On the Planning Abilities of Large Language Models."

53. Eric Horvitz, "Principles and Applications of Continual Computation," *Artificial Intelligence* 126, no. 1–2 (February 2001): 159–196, https://doi.org/10.1016/s0004-3702(00)00082-5.

54. Nicholas Roy et al., "From Machine Learning to Robotics: Challenges and Opportunities for Embodied Intelligence," *arXiv*, October 28, 2021, https://doi.org/10.48550/arXiv
.2110.15245.

55. T. Mitchell et al., "Never-Ending Learning," *Communications of the ACM* 61, no. 5
(April 24, 2018): 103–115, https://doi.org/10.1145/3191513.

56. Suriya Gunasekar et al., "Textbooks Are All You Need," *arXiv*, October 2, 2023,
https://doi.org/10.48550/arXiv.2306.11644.

57. Sébastien Bubeck and Mark Sellke, "A Universal Law of Robustness via Isoperime-
try," *Journal of the ACM* 70, no. 2 (March 21, 2023): 1–18, https://doi.org/10.1145/3578580.

58. Gunasekar et al., "Textbooks Are All You Need"; Arindam Mitra et al., "Orca 2:
Teaching Small Language Models How to Reason," *arXiv*, November 21, 2023, https://doi.org
/10.48550/arXiv.2311.11045.

59. Saleema Amershi et al., "Guidelines for Human-AI Interaction," *Proceedings of the
2019 CHI Conference on Human Factors in Computing Systems* (ACM, 2019), https://doi.org
/10.1145/3290605.3300233; Eric Horvitz, "Principles of Mixed-Initiative User Interfaces," *Proceedings of the SIGCHI conference on Human factors in computing systems the CHI is the
limit—CHI '99* (ACM, 1999), https://doi.org/10.1145/302979.303030.

60. Abigail Sellen and Eric Horvitz, "The Rise of the AI Co-Pilot: Lessons for Design from Aviation and Beyond," *Communications of the ACM* 67, no. 7 (June 13, 2024): 18–23, https://doi.org/10.1145/3637865.

61. Mitra et al., "Orca 2."

62. Gunasekar et al., "Textbooks Are All You Need."

63. Qingyun Wu, Gagan Bansal, Jieyu Zhang, Yiran Wu, et al., "AutoGen: Enabling Next-Gen LLM Applications via Multi-Agent Conversation," paper presented at the Conference on Language Modeling, Philadelphia, PA, October 7–9, 2024.

64. John Jumper et al., "Highly Accurate Protein Structure Prediction with AlphaFold," *Nature* 596 (2021): 583–589, https://www.nature.com/articles/s41586-021-03819-2.

65. Minkyung Baek et al., "Accurate Prediction of Protein Structures and Interactions Using a Three-Track Neural Network," *Science* 373, no. 6557 (August 19, 2021): 871–876, https://www.science.org/doi/10.1126/science.abj8754.

66. Ian R Humphreys et al., "Computed Structures of Core Eukaryotic Protein Complexes," *Science* 374, no. 6573 (December 10, 2021): eabm4805–eabm4805, https://doi.org/10.1126/science.abm4805.

67. Joseph L Watson et al., "De Novo Design of Protein Structure and Function with RF-diffusion," *Nature* 620, no. 7976 (August 2023): 1089–1100, https://doi.org/10.1038/s41586-023-06415-8; Wu et al., "AutoGen."

68. Microsoft Research AI4Science and Microsoft Azure Quantum, *The Impact of Large Language Models on Scientific Discovery: A Preliminary Study Using GPT-4* (November 2023), https://arxiv.org/pdf/2311.07361.

69. Eric Horvitz et al., "Now, Later, and Lasting: Ten Priorities for AI Research, Policy, and Practice," *Communications of the ACM* 67, no. 6 (June 1, 2024): 39–40, https://dl.acm.org/doi/pdf/10.1145/3637866.

70. Joseph R. Biden, "Executive Order 14110: Executive Order on the Safe, Secure, and Trustworthy Development and Use of Artificial Intelligence," *Federal Register* 88, no. 210 (October 30, 2023): 75191–75226.

Perspectives on AI from Across
the Disciplines

NAS/NAE/NAM Working Group Members

Authored by the working group participants who are members of the National Academy of Sciences (NAS), the National Academy of Engineering (NAE), or the National Academy of Medicine (NAS), the perspective pieces included in this chapter responded to the conveners' request to briefly describe how AI had or could transform their disciplines. We include an edited digest of the responses they delivered at the two Sunnylands convenings. Academy members are listed in alphabetical order followed by their respective NAS, NAE, or NAM section(s).

—Kathleen Hall Jamieson, William Kearney, and Anne-Marie Mazza

David Baltimore (NAS, NAM)

Microbial Biology
Biochemistry, Cellular and Developmental Biology, Medical
Microbiology and Immunology, and Genetics

Biology has been undergoing a continual revolution since I began working in biology in 1960. And the depth of that revolution, call it every five years, is astounding. The things that we think about today and the things that we do today bear no relation to what we did five years ago, ten years ago, 20 years ago. We've had to adapt to this continuing revolutionary behavior because it's exciting and because each revolution generates a new depth of understanding.

Now the latest involves melding AI into what we built over these many years, and the results have been astounding. AlphaFold, which allows us to predict fairly accurately the structure of proteins from simple sequence, something we dreamed of doing, is itself one of these revolutionary moments. So we in biological research have thought a lot about how you control something moving at this extraordinary pace. Most of you will be aware of the Asilomar process of many years ago, where we established a procedure for taking on a revolutionary methodology that might have downsides as well as upsides. That set the stage for the more recent adaptation of genome editing into our portfolio of techniques, the potential of changing the inheritance of our whole race. I hope that thinking about these precedents will be in the agenda of this meeting, and biologists can be a little helpful with the experience that we've had in thinking about regulatory issues and the other meta issues that allow science to move forward.

Vinton G. Cerf (NAS/NAE)

Computer and Information Sciences
Computer Science and Engineering

In 1943 Warren McCullough and Walter Pitts invented the perceptron. In 1957, Frank Rosenblatt implemented the concept using a neural network that had three layers and was capable of classifying groups of objects that are linearly separable (e.g., by drawing straight lines in a two-dimensional space). More complex separation functions required more layers, as researchers discovered in subsequent work.

The term *artificial intelligence* (AI) was coined at a time when computers were still relatively new (early 1960s) and had been called "artificial brains" by some. There was a kind of hubris in this nomenclature. Researchers began with heuristic programs that seemed to exhibit intelligent behavior some of the time. A more codified version of AI emerged, called expert systems, that used a programming structure based on if-then-else logic. For example, "if this symptom is present, then there is x probability of some diagnosis, else check for a different symptom."

In the 1980s, some researchers returned to multilayer neural networks to solve more complex classification problems. The primary computations required to adjust the weights of each neuron are forms of matrix multiplication. With the repurposing of graphical processing units, whose original purpose was fast triangular texture computation, to do fast matrix multiplies, neural networks became capable of surprisingly effective applications including speech recognition, translation, image classification and, ultimately, text and multimodal synthesis in response to a prompt. At Google, even more specialized hardware, called tensor processing units (TPUs), has been developed.

This generative form of AI has become topic A in the last few years, leading to a great deal of hyperbolic speculation about the capability of these specialized computing systems we now call large language models (LLMs). Some of them can write software. Others can distinguish and classify images. Others can synthesize voices, videos, and images from text prompts with considerable creativity. It is not hard to become excited to see systems like ChatGPT producing poetry like this haiku, which resulted from a prompt: "write a haiku about a rose in Shakespearean style":

> *ChatGPT:* In fairest garden, Sweet rose blooms 'neath summer's
> gaze, Beauty's fragrant blush.

This may be compared to one of Shakespeare's related sonnets written in 1609:

> From fairest creatures we desire increase, That thereby beauty's
> rose might never die, But as the riper should by time decease,
> His tender heir might bear his memory:
> But thou, contracted to thine own bright eyes, Feed'st thy light's
> flame with self-substantial fuel, Making a famine where
> abundance lies,

Thyself thy foe, to thy sweet self too cruel:
Thou that art now the world's fresh ornament, And only herald to
    the gaudy spring,
Within thine own bud buriest thy content,
And tender churl mak'st waste in niggarding: Pity the world, or
    else this glutton be,
To eat the world's due, by the grave and thee.[1]

There are many much more compelling examples that can be offered. Some of these LLMs can produce software. Others generate text, sound, imagery, or video based on text prompts. Others write essays or provide advice in response to queries.

One might reasonably wonder, "Why are these artifacts so seemingly creative, knowledgeable, and intelligent?" The training of the neural networks involves the ingestion of large quantities of text that has been tokenized. A token can be a word or a phrase. A high dimensional model is created to capture the probability of a word occurring after the input of a prompt or a line of text. During the training period, an LLM is formed using random weights associated with the neurons of a multilayer neural network. The model starts out with a fairly poor representation of the statistical relationships among the many (sometimes billions) of tokens. The training consists of presenting the model with partial sentences, asking the model to "fill in the blanks." In what is called back-propagation, good responses, for some value of good, adjust the producing weights accordingly. Bad responses cause the weights to be adjusted to be less likely to produce the undesired response.

A text LLM is essentially a statistical reflection of the text that it was trained on. It could be thought of as a compressed representation of the ingested text. It should not be surprising that this statistical representation produces conceptually coherent output that mimics human discourse and even reasoning. After all, the content used to train the model had meaning that was expressed more or less coherently. The statistical model of which words might reasonably follow preceding words contains some of that knowledge. If the source material is grammatically correct and uses a broad vocabulary of words, it can mimic human discourse in convincing ways.

It is precisely because of this convincing mimicry that one is led to imagine that the LLM (or "bot") is nearly sentient. Of course, it is nothing of the sort. It is simply a generative system that is driven in part by the coherent expression of fact and belief contained in the training material.

Because of its statistical nature, a generative LLM can also produce counterfactual output in response to prompts. The training often lacks any indication of context such that text can be generated that draws on words occurring in different contexts that, when strung together, produce false assertions. This phenomenon is sometimes referred to as hallucination. We are still some distance from understanding how to curb this tendency.

Perhaps even more disturbing is that these generative systems produced generally very good quality sentences. These often sound very anthropomorphic: "I am just a chatbot." The self-reference imbues the system with the verisimilitude of humanity and self-consciousness. Users of these systems sometimes see the responses as empathetic and often give them credibility for social awareness that the LLM does not deserve. Of course it sounds humanistic; it was derived from the expression of human discourse and writing!

These systems produce the illusion of human discourse and are often extremely convincing, even when completely wrong. We are learning to use them in myriad ways but should be wary of being misled by the glib responses to our prompts. Critical thinking is our friend in the use of these artifacts. At some point, perhaps there will also be norms, rules, and even regulations to protect users from taking advice that sounds authoritative but is dangerously wrong. We have a lot to learn about these complex artifacts and meanwhile should be wary of their application for anything that might be high risk.

Joseph S. Francisco (NAS)

Chemistry

In the field of chemistry, AI is being increasingly embraced by both publishers and researchers. Its application has surged significantly in recent years, revolutionizing various aspects of journal operations. AI is now instrumental in enhancing manuscript quality by aiding authors in refining their work with greater precision. Moreover, AI's role extends into the production process, a development that many, including myself, were previously unaware of. However, this rapid integration of AI also brings with it a few growing concerns.

One major concern is the role of AI in aiding paper mills. AI can help paper mills produce papers that evade detection, making it challenging for journals to identify submissions originating from paper mills. The papers produced by paper mills often lack real data, have manipulated images, and have authors without institutional emails or public records, who are hard to trace.

AI tools are being used by publishers to detect paper mill submissions by reviewing visual content and sub-images to identify discrepancies. Such tools can flag duplicated and manipulated images and figures before publication, enabling publishers to correct unintentional errors or reject manipulated manuscripts. Publishers are actively working to detect these kinds of submissions to maintain the integrity of the research they publish in their journals.

In chemistry, analytical chemistry and biochemistry lead in integrating AI into their research compared to other subfields. However, areas like organic synthetic chemistry have not yet seen AI's influence. Despite this, the field holds tremendous potential for AI to facilitate the discovery of new molecules.

An emerging issue in organic chemistry involves the use of AI to generate synthetic procedures. The question is whether AI can reliably produce procedures to synthesize molecules. Additionally, if a novice chemist follows an AI-generated procedure without adequate experience, it could lead to dangerous situations. To mitigate this risk, we need to ask upfront questions: Should there be post hoc filtering for the synthesizability of AI-generated results? If a procedure is generated by AI without validation, who is held accountable—the AI, the user, or the journal? This presents significant chemical safety issues that have not yet been fully addressed.

AI systems are typically trained on representative datasets, learning from these datasets to formulate predictions based on observed patterns or to generate new data. Consequently, AI models require accurate and readily accessible datasets. However, the reliability of the databases used to train AI remains a significant concern. Many databases that AI systems rely on lack reliability, even though some dependable databases do exist. The success rate of various databases and libraries used to train AI is not well established. An important benchmark in this context is the number of synthetic steps required to produce molecules generated by AI. This emerging issue in synthetic chemistry might explain why fewer researchers are integrating AI into this area.

Barbara J. Grosz (NAE)

Computer Science and Engineering

Generative AI models have been changing computer science in the various ways John Hennessy describes for engineering fields at large, and their ability to help researchers find relevant prior work has the advantages and challenges he notes. More profoundly, generative models are providing new ways of interacting with computer systems, and, as they have proven capable of producing useful segments of code, they are radically changing the ways programmers work. Computer science education is changing as a result.

AI research in natural-language processing has had as its goals understanding people's linguistic capacities and building systems that could match those capacities. It aimed, in part, to enable systems to participate in dialogues similar to those that occur when people talk with one another. Generative AI methods have enabled stunning successes in natural-language processing. In myriad ways, dialogues with systems based on these models now help people more easily use computer systems to find information and accomplish tasks across a broad range of domains. Though carried out in the languages people ordinarily speak, and thus more natural than programming, these dialogues lack certain features of human-to-human dialogue. A kind of guided, sometimes collaborative, search for an answer, suggestion, or solution, they are a mixture that is best captured by the new phrase (and job opportunity), *prompt engineering*: the prompts are in a natural language, but the need for engineering is a symptom of the distinction.

The engineering must be good engineering; for that, anyone who does prompt engineering needs to learn not only effectiveness and efficiency but also ways to judge the quality of the results. Years ago, a theoretical physicist railed at me that computing was making his students less competent modelers. They "just coded," without questioning the answers they got back; they trusted the computer and had not developed intuitions that immediately made them consider whether the answers it generated made sense. For any complex computing system, it is hard to know whether a program does what one intends and expects it to do. Our current inabilities to understand why generative models produce the answers they do, and the hallucinations for which they are well known, exacerbates the problem of knowing whether the code they produce actually correctly performs the functions a user intends. Computer scientists are as susceptible to pro-automation bias—the assumption that if a computer produces an answer, it is right—as others. Generative AI thus raises a critical new challenge for AI researchers, that of verifying the results that prompt dialogues produce. Meeting this challenge is likely to require expertise from several other areas of computer science, for instance, work on program verification and on interaction techniques. Notably, the new methods that are developed could be useful far beyond programming and computer science.

Computer science education is evolving in light of these new AI capabilities. Students, like professional programmers, are now using generative AI systems to code for them. They differ from professional programmers, though, in the amount of experience they have programming without such support. How will they develop intuitions for detecting if the AI system has provided good code? What new skills do they need to learn for debugging and testing? The powers that generative AI has released makes it even more important for computing researchers, developers, and systems' deployers to consider not only what systems they could build, but what systems they should build and the right way to design them, taking account of their potential impacts on communities and societies as well as individuals. Teaching the skills to reason about such matters is also beginning to become part of computer science education in some institutions.

John L. Hennessy (NAS, NAE)

Computer and Information Sciences
Computer Science and Engineering

We find ourselves in an interesting and fast-moving era. As Eric Horvitz and Tom Mitchell have discussed in their survey chapter, the emergence of deep leaning has created a discontinuity in the capabilities of AI systems. Many new engineering faculty members across a wide variety of disciplines have machine learning in a description of their research. We are seeing an incredible revolution in engineering in which these machine learning techniques are going to be used as scouts to find novel approaches to problems and as tools to narrow the solution space, particularly for complex, high-dimension optimization problems.

For example, researchers exploring new battery structures might use deep-learning techniques to search for materials that avoid some of the downsides of lithium. Another researcher might explore new methods to capture methane. One of the most amazing applications I have seen is the use of machine learning to understand turbulence and turbulent flow. Turbulence is a classically hard problem that has resisted most of our numerical attempts. A breakthrough in analyzing and understanding turbulent flow would have applications in the design of wind turbines, automobiles, trains, and planes, as well as applications in other areas.

Of course, these deep-learning systems can be joined with traditional computational methods, as AlphaFold does. AlphaFold isn't just an AI system. It uses computational techniques as well. Melding these techniques together, allows a researcher to combine the strengths of each. The deep-learning system may work better to determine the overall structure of a protein, while computational techniques may be more useful at fine-tuning the structure.

For engineers, finding the general structure of a solution is only step one of a process to realize a product that efficiently solves a real problem, which in the end is what drives engineers. Of course, you must worry whether an AI system is guiding the researchers in the right direction. Just as in other applications of machine learning, verification of the accuracy of predictions will be important, and that will likely require human intelligence for some time to come.

Eric Horvitz (NAE)

Computer Science and Engineering

Reflecting on the current state of AI, I find myself immersed in two inter-related realms: the *scientific advancements* of AI and their *societal impacts*. We are in an exciting period for AI, with the capabilities of neural network models rising faster than our understanding of the principles underlying the emergent behaviors we are observing. These advancements have stim-ulated scientific curiosity and catalyzed new directions for AI research, bringing novel questions, methods, energy, and intensity to colleagues and teams that I collaborate with. Simultaneously, the rapid diffusion of AI tools into everyday life has deepened my sense of responsibility regarding the short- and long-term societal influences of AI technologies. I have invested increasing time and resources in reflecting on and addressing potential dis-ruptions, ethical concerns, and the opportunities AI presents in various realms.

Scientific Journey

I was drawn to do my doctoral work in AI as the best path forward for gaining an understanding of the mysteries of human cognition. Close colleagues and I contributed to the ignition of a probabilistic revolution in AI, moving away from the dominant logic-based methods of the time, and working to advance the development of AI systems based on a founda-tion of probability and utility theory. The axioms of probability were ex-tended in the 1940s to inferences about taking ideal actions in the world via the axioms of utility theory, as first formulated by von Neumann and Morgenstern. Probability and utility theories form a widely assumed and celebrated set of principles that have been considered a *normative basis* for rational reasoning and decision-making. There are multiple challenges with building AI systems in accordance with these principles, including computational complexity. A long-term complaint in AI was that the normative basis was unrealistic in terms of the requirements for computational resources. I focused during my doctoral efforts and for many years later on developing principles and models of *bounded rationality* built on a foundation of probability and utility that could enable systems with limited computational resources to perform well amid the complexity of the open world. The work included the development of formal mechanisms for guiding evidence gathering and inference. Other teams explored numerous other approaches for leveraging probability in representations and reasoning. This shift to a rationalist approach to AI—harnessing a normative foundation of probability and utility—became central in advancing machine learning, perception, reasoning, and decision-making.[2] The approach enabled the community to build systems that could address real-world challenges, such as providing recommendations on medical diagnoses and decisions. The rationalist approach provided clear semantics and a strong theoretical foundation for building systems operating on understandable and sound principles.

Recent advancements in neural network models mark a significant inflection point in AI's trajectory.[3] Impressive capabilities and rates of improvement are seen in vision, speech recognition, and language understanding benchmarks. Generative AI has recently emerged with models being built at increasing scales demonstrating surprising powers in generating language, images, video, and molecules. Neural-network models are being harnessed in numerous areas, including the sciences. For example, advances in predicting protein structure and drug design are accelerating research in the biosciences, including efforts to design new therapeutics.

Despite the excitement, we grapple with the relationship of neural models to prior advances. In distinction to the clarity of previous work based on the rationalist approach, much of the detailed operation of generative models remains a mystery. Neural networks have thrust us into empirical studies of these large-scale systems, akin to methodologies for probing and studying nervous systems.[4] This jump, from a successful multidecade trajectory of advances with rationalist approaches in AI to the mysteries of neural networks, frames intriguing and interesting opportunities to pursue answers to significant questions that remain unanswered. We face a critical scientific challenge of bridging the gap between empirical observations of the behavior of neural networks and foundational principles of well-understood theories of inference and action. I hope to see bridges constructed over the next decade.

Societal Implications and Responsibilities

I believe that AI scientists and engineers have a critically important role and responsibility to identify and share technical developments that have implications for people—and society more broadly. Responsibilities include informing and engaging with multiple stakeholders across domains and sectors and working to broaden awareness and participation. This work involves being available for expert consultations, organizing and participating in special meetings and engagements around milestone developments, and establishing committees, organizations, and initiatives for tracking, guiding, and communicating about AI advances over time.

Fifteen years ago, AI was beginning to make its way into real-world applications as I assumed the presidency of the Association for the Advancement of Artificial Intelligence (AAAI). I themed my presidency "AI in the Open World," highlighting the need to develop AI systems that could perform robustly and in a trustworthy manner on real-world tasks, and also our responsibility to understand and address the potential societal impacts of the AI systems that we build.[5] To explore societal influences, I commissioned the AAAI Presidential Panel on Long-Term AI Futures. This initiative culminated in a retreat at Asilomar in 2009, chosen for its symbolic connection to the historic meeting on recombinant DNA.[6] The clear value of the discussions and collaborations at the AAAI Asilomar retreat and premeetings sparked the establishment five years later of the One Hundred

Year Study on AI at Stanford, which was created to bring experts together every five years to observe, synthesize, and provide assessments and guidance in the spirit of the AAAI Asilomar meeting.[7] The study is endowed to continue this process for the life span of Stanford University. Projects of the study include the creation of faster-paced analyses, including the AI Index, an annual assessment of AI capabilities and influences.[8]

Beyond recurrent studies by experts, the ubiquity of AI's influences requires that diverse voices participate and help to guide the development and use of AI systems. AI scientists have a responsibility to organize, alert, and educate a spectrum of stakeholders—as well as to establish venues for listening and responding. In 2016, AI scientists from industry, academia, and nonprofit research centers cofounded the Partnership on AI, bringing together stakeholders from industry, academia, and civil society to foster discussions, analyses, and make recommendations on the responsible advancement of AI.[9] As the founding chair, I've observed the power of bringing scientists together with policymakers, civil liberties experts, and a broad spectrum of civil society organizations. While still in its first decade, the Partnership on AI has already made significant contributions to multiparty collaboration on key topics.

With potential fast-paced developments, AI scientists may need to engage quickly at times and bring diverse expertise to the table as early as possible when new capabilities and issues arise. Given the behaviors I saw in our internal studies of an early prerelease version of GPT-4 in August of 2022, I felt it important to gain permission to share the confidential prerelease model with experts across disciplines. This initiative led to the *AI Anthology* effort, which provides multiple viewpoints on how the new capabilities might be best leveraged for human flourishing.[10]

AI scientists need also to inform and provide guidance to government agencies and leaders about technical advancements with AI and work with policymakers on steps forward. It has been an honor to be invited to testify on AI at both open hearings and closed sessions of Congress[11]—and to have opportunities to engage with senior leadership at the White House and colleagues via my role as a member of the President's Council of Advisors on Science and Technology (PCAST).

These diverse projects, engagements, and organizational efforts are examples of AI scientists' responsibilities to engage and inform across sectors, to work to broaden awareness and participation, and to promote research on AI's responsibilities, ensuring that we include multiple voices in assessments and decisions, and that we stay ahead of the innovation wave with technical, sociotechnical, and regulatory advancements.

Moving Forward

Looking ahead, the interplay between AI's scientific advancements and societal impacts will become even more critical. We urgently need to grow our scientific understanding of the operation of systems built on neural network methodologies. Better scientific understandings will help us to shape the development and application of safe, reliable, fair, and understandable AI methods. We need to complement curiosity-driven research and the thrill of scientific breakthroughs in AI foundations with investments in technology and policy to understand, shape, and regulate influences of the technologies on people and society. This work includes ongoing study spanning technology, design, and psychology of human-AI interaction.[12]

The potential benefits of AI are immense—from accelerating scientific discovery to improving education and raising the quality of health care outcomes. However, we have to consider recognized risks, particularly with information and media integrity, biosecurity, fairness and equity, safety and reliability, and privacy and security. We must also stay on top of "deep currents" of more complex interactions of AI with culture and society, such as how these systems may change and disrupt—in costly and in valuable ways—education, the creative arts, scientific discovery, jobs, and the economy. We must work to monitor and come to better understandings of the subtle but potentially powerful influences of AI applications on the human psyche, including the impacts on our human dignity and agency.[13] Outcomes need not be dominated by situations and equilibria reached via laissez-fair flows of technology into society. With the maturation of AI and its applications, we have opportunities to manage and guide the technology with foresight and responsibility.

The current state of AI is marked by fast-paced progress and significant challenges. As a scientist driven by curiosity about human cognition and devoted to reaching understandings of computational principles of intelligence, I'm excited by potential AI discoveries, machinery, and new applications on the horizon. At the same time, I am cautious and concerned about the influences of AI innovations on people and society. We need to make investments in steering AI's development to promote human well-being and societal progress. Through continued scientific exploration and a thoughtful, inclusive, and multidisciplinary approach to applications and influences, we can leverage AI as a force for good, advancing our understandings of the scientific foundations of intelligence and enriching human society. AI scientists, with their unique insights, must lead at the frontier, providing awareness of developments and implications and a commitment to engage with the public, civil society organizations, government leaders and agencies, and experts across various fields to address these responsibilities and to help shape AI's future.

Kathleen Hall Jamieson (NAS)

Social and Political Sciences

Three of our retreatants—Barbara Grosz, Mary Gray, and John Hennessy—played important roles in shaping the germinal National Academies of Sciences, Engineering, and Medicine report, "Fostering Responsible Computing Research: Foundations and Practices," that grounds our deliberations. By observing that "The social and behavioral sciences provide methods for identifying the morally relevant actors, environments, and interactions in a sociotechnical system," that report draws attention to the role that the social and behavioral sciences should play in framing discussions and decision-making about generative AI.[14]

It is the behavioral and social sciences, for example, that remind us that our language and our frames embed assumptions about ethics and equity about which we are largely unaware, a point made in Chapter 7 by Shobita Parthasarathy and Jared Katzman. Raw data for example are not "raw" but rather the product of choices and the values of those who frame the research questions, privilege some methods over others, and in the process determine what is and is not considered evidence and proof. At the same time, conventionalizing the language of "artificial intelligence" risks changing our sense of what it means to say that someone or something is intelligent.

Social scientists who focus on human interaction and the ways in which humans act within social and political structures are grappling with such questions as: How does what we humans know, how we know it, and how we interact with each other and make sense of our worlds change if AI is layered atop the dispositions that humans have to deceive, distort, and act on their fears and venal impulses?

FactCheck.org, which I cofounded, was premised on the idea that journalists could arbitrate disputes about "fact" by turning to evidence in impartial trusted sources such as the Bureau of Labor Statistics and the National Academies that honor scientific norms and have generated reliable knowledge in the past. That common knowledge could in turn help ground deliberation and governance. However, in an AI world, someone who seeks out the National Academy of Sciences' website may find a hyperrealistic but fake site featuring a supposed President Marcia McNutt, who looks, sounds, and seems more like Marcia than Marcia herself but is promulgating pseudo-science. How can factcheckers or the public tell that the deepfake is not deep reality? I wrote a book on how Russian trolls and hackers helped elect a president in 2016. If you add the currently available AI technologies to their equation, the Russians would have succeeded to an even greater extent because their efforts probably would have gone undetected.

As AI scientists are developing ways to identify and constrain AI-generated content, social and behavioral scientists are among those probing its impact both for good and ill on democratic systems and informed voting as well as on how and what we know and how we interact with each other and with these new technologies.

Marcia K. McNutt (NAS, NAE)

Geology
Earth Resources Engineering

In my view, there is hardly a field that has more benefited from AI, but also is imperiled because of AI, than the environmental sciences. And the benefits have come because so much of this universe is inaccessible to humans or only accessed by humans at great cost and peril.

Deep space exploration was one of the earliest applications of an AI precursor called "automated planning and scheduling." These smart systems used sensors on space probes to allow an unmanned vehicle itself to make decisions on operations based on what it was learning from its own instrumentation without having to endure the delay in sending data back to Earth for a human to make the decision.

Deep sea exploration followed suit and delivered even a higher payoff application. While underwater exploration can be conducted directly by humans, it is only with much sacrifice. Deep-diving human-occupied submersibles are cold, cramped, and uncomfortable for any length of time, and their use is further limited by high cost and extremely limited range. Remotely operated vehicles are more affordable but require an umbilical-cord tether to provide power from a surface ship and control from a ship-based pilot because the ocean is opaque to electronic message transmission. However, the tether restricts the spatial extent of the mission. Conversely, automated planning and scheduling installed in autonomous (untethered) underwater vehicles totally revolutionized our opportunities to explore the deep sea, both in cost and complexity of the mission. AI-guided vehicles can make their own decisions, execute complex search patterns in all dimensions, collect data and samples, know when the mission is accomplished, and then return home loaded with data and samples. Humans no longer needed to be involved in real time. AI-guided autonomous vehicles have greatly reduced the cost of exploration of hostile environments and increased the scientific return.

These systems were likely the forerunners of today's automated driving routines, except that there was no safety issue in the deep space or ocean.

If the vehicle misidentified something and ran into it, no one was going to die the way it is with cars on congested roads.

Other areas of the environmental sciences are benefiting from AI beyond exploration of Earth and space. For meteorology, AI is able to forecast more accurate weather predictions and track dangerous storms like hurricanes. AI could as well predict the impact of some interventions on climate change. Using the same advances that allow AI systems to distinguish faces, AI is now regularly used to identify plants, animals, and other natural features from photos. This capability has been a boon to citizen science, for example, in improving the accuracy of annual bird counts.

On the negative side, I am concerned about the impacts to the environmental sciences from very successful and convincing fakeries, especially in terms of climate science. So much is at stake with our response to the current climate crisis that big money will be invested in trying to debunk climate science and in arguing that interventions are not worthwhile. Our ability to detect when AI has been used in malevolent ways to overturn what is strong scientific consensus is constantly being challenged by more convincing fakes.

Saul Perlmutter (NAS)

Physics

In physics, cosmology, and astrophysics, some of the more frequent AI applications that we've seen have to do with speeding up simulations. Simulations have become a large part of so much science nowadays. It makes it possible to hunt for rare solution spaces that you wouldn't have ever considered with a slow simulation. You might now be able to hunt for those solutions with a fast mimic of the simulation that you get with AI.

This also means that AI changes a lot about how we do statistics. Over the years, we've moved toward more and more Monte Carlo–style statistics where you mimic the system that you're working with, do many renditions of it, and that gives you the contours of statistics, rather than calculate them from first principles. This is another real advantage of having a much faster technique for simulation. Statistics plays a huge role in the sciences, one that we don't usually talk about (unless statistics is your field!). It's a hidden-in-plain-sight important tool, and I think it's going to change dramatically with this AI capability.

Large language models also can enable advances that combine different fields, because they make it easier to do cross-disciplinary translations. I've already found myself in meetings with people from different fields during which I quickly looked up terminologies, acronyms, and jargon that they're using in those fields, and this allowed me to be part of the cross-disciplinary conversation within a matter of seconds. Previously, you would have to go back and find this whole body of knowledge. That's going to be an important game changer since so many scientific developments have to do with working at the edges between different domains in different fields.

Similarly, AI offers more fluid data wrangling. So much data science involves getting data from point A to point B in a form that you can use, and we're finding that these AI systems are very helpful in making that possible. You can read entire datasets without looking up the manuals. AI can explain to you what every column is, and it actually does a good job in giving you a structure to be able to work with data that you might otherwise never have accessed.

Finally, people can take more mathematically sophisticated approaches because you can treat entire mathematical derivations as if it's a calculator helping you do an arithmetic problem. And so that makes some activities much faster.

I don't yet know about the idea of using AI to stimulate new ideas, like feeding the AI a bunch of papers and asking it "What's missing here?" I can't tell yet whether this is already something that's becoming useful or whether it's something that we might expect to become useful in the next generation. Here, a big concern is that we don't want to get into idea feedback loops where the AI is training on material that comes out of people working with the previous generation of AI. We want to make sure that we don't inadvertently feed our AI-generated material back into the AI training.

William H. Press (NAS)

Computer and Information Sciences

AI will be transformative, but I am not waiting for the transformation. Right now, I use it every day for quick facts and a range of administrative and programming tasks that I would once have characterized as frustrating, fussy, or boring. (Of course, I always check the results.) Here are a few of my recent prompts:

1. "From various journals I have cut and pasted a bunch of references below for a paper I am writing. Please convert them all to PNAS format."

2. "Where did the funding of the Einstein Foundation in Berlin originally come from? I want to be sure that it is not money from a controversial industrial source. Please check your answer against reputable Web sources."

3. "What serious human diseases are thought to already have been endemic in the Native American population in pre-Columbian times?"

4. "Give specific names of good reviewers for a paper that builds a large NN model (not an LLM) for predicting results from a large combinatorial biology experiment? It's similar to, but different from, drug-discovery, so I want people with broader ML and NN experience. I especially want names of junior faculty at good universities."

5. "In Python with Numpy, if I write something like neww = oldd[3:6,10:15], does neww point to data within oldd, or is a copy made?"

6. "I have a Jupyter notebook named mynotebook.ipynb . In Python, how can I extract the text of a particular cell and then reformat it to LaTeX format? The cell I want begins with the comment #ThisCellPlease."

7. "In Python with Pandas and Numpy, I have a dataframe df. All entries are small integers. I want to make a large crosstabulation where each column is expanded to its number of unique values.

So, the crosstabulation will be an N by N matrix where N is the sum of the number of uniques for each column. How do I do this? Code only, please, no explanations."

8. "I have a very big numerical dataframe and want to fit it with a Gaussian copula, and then generate synthetic rows from the fitted model. Show me PyTorch code for doing this efficiently on a CUDA GPU."

9. "I have an HTML and PHP page that uses Google's Recaptcha v2 like this:

```
$recaptcha = $_POST["g-recaptcha-response"];
$secret_key = 'my-secret-key';
$url = 'https://www.google.com/recaptcha/api/siteverify
    ?secret='
. $secret_key . '&response=' . $recaptcha;
$response = file_get_contents($url);
$response = json_decode($response);
```

What would the code be to upgrade this to Google's Recaptcha v3?"

Jeannette M. Wing (NAE)

Computer Science and Engineering

We are witnessing unfettered growth in the deployment of AI systems in critical domains such as autonomous vehicles, criminal justice, education, health care, and public safety, where decisions taken by AI agents directly impact human lives. This growth underlines the need for computer scientists to understand and harness this technology better.

We need a scientific understanding of why today's AI models work so well. We do not know their mathematical properties. We do not know how to quantify or predict their behavior. We do not know how to explain why an AI model produces one result and not another. Small perturbations to input data can lead to wildly different outcomes. When will adding more compute and more data to build larger models hit a wall? Experimentation in AI is far ahead of any kind of theoretical understanding.

We need trustworthy AI.[15] How can we trust decisions made by AI models to be accurate, fair, reliable, robust, safe, and secure, especially under adversarial attack? One approach is to use formal methods, based on mathematical logics and symbolic reasoning, to provide provable guarantees about AI systems. Formal methods applied to AI would require probabilistic reasoning and characterizing verifiable properties of real-world data.

AI raises new ethical issues. The Belmont Principles of beneficence, justice, and respect for persons are a good starting point for AI. They need to be lifted to operate on groups of individuals, not only on individuals. Finally, we need to revisit the codes of conduct in all professions that incorporate the use of AI.

Michael Witherell (NAS)

Physics

I am speaking as a leader of Berkeley Lab, where I have the privilege of leading 1,600 scientists working in a wide range of science and technology. As part of my job, I've had the joy of reading published impactful research on applying machine learning techniques in cosmology; particle and nuclear physics; material science; synthetic biology; matter genomics; environmental biology; geoscience; climate modeling; and smart grid, water treatment, and accelerator operations.

AI has had a transformative effect across all these fields of science, but much of the effort is invested in developing stable, robust, interpretable methods that can be explained, exposed, and verified to a skeptical scrutiny of researchers in these fields. And that's actually what a lot of the work has been. What is often the primary barrier to accelerating R&D using AI is not the computing power available but rather the size and quality of the experimental or computational datasets available for training the models.

As an example, most of the data on local ecosystems were collected in small projects, producing specialized datasets in many areas that will not take advantage of meta-analysis in general, let alone AI, unless we have interoperable datasets. Several groups around the country are working on projects to integrate these datasets, including one at Berkeley Lab.

I would like to offer another quick example that has been recently in the news. Researchers have developed fast, agile, and reliable weather models using AI that offer an unprecedented level of high-resolution information. Such models could provide improved guidance to prepare communities for extreme weather events. In a very short time, these models have gotten to the point that their results are as reliable as for the traditional models and can be run much faster. The new models produce a range of scenarios, each one taking less than two seconds, which is several orders of magnitude faster than existing models. One can now create huge ensembles of predicted weather outcomes, greatly increasing the ability to forecast low-probability, high impact events. Consider the lives that could be saved if such models can be made very reliable and if the predictions they make can be communicated in a way that is trusted by the public.

Most of the AI-enabled advances in research to date have been accomplished with special purpose models. Because the remarkable general-purpose large AI models are so new, we still need to understand their full potential to accelerate scientific research. If we consider the fields of science in which the data is not personal data, the principal risk is that an apparent discovery might be due to an artifact or a hallucination by the model. How do you show that this new type of black hole is real and not something that was manufactured by the model? One must closely embed computer scientists with physical scientists, biologists, and climate scientists from the beginning of the research project. By working together as an integrated team they can develop analytic tools that will produce verifiable results able to stand up to rigorous scrutiny by the scientific community.

Finally, although many of these areas do not work with human data, they still can have complex and sensitive interactions that have ethical and societal implications. For example, consider a system for detecting, measuring, and reporting methane leaks using satellite data and ground-based observations, all integrated with AI. This is a really important problem with great significance for the global community. Who is to be trusted with the design and use of such a system? An oil and gas company, a consortium of utilities, the US Department of Energy? The governance of such a system is critical in making sure it serves all of us well.

Notes

1. William Shakespeare, Sonnet 1, 1609.

2. John von Neumann and Oskar Morgenstern, *Theory of Games and Economic Behavior* (Princeton, NJ: Princeton University Press, 1944).

3. Sébastien Bubeck, Varun Chandrasekaran, Ronen Eldan, Johannes Gehrke, et al., "Sparks of Artificial General Intelligence: Early Experiments with GPT-4," *arXiv* preprint arXiv:2303.12712, March 22, 2023, https://arxiv.org/abs/2303.12712.

4. Kevin Meng, David Bau, Alex Andonian, and Yonatan Belinkov, "Locating and Editing Factual Associations in GPT," *Advances in Neural Information Processing Systems* (2022), 17359–17372; Catherine Olsson, Nelson Elhage, Neel Nanda, Nicholas Joseph, et al., "In-Context Learning and Induction Heads," arXiv:2209.11895, September 24, 2022, https://arxiv.org/abs/2209.11895; Mert Yuksekgonul, Varun Chandrasekaran, Erik Jones, Suriya Gunasekar, et al., "Attention Satisfies: A Constraint-Satisfaction Lens on Factual Errors of Language Models," *arXiv* (2023).

5. Eric Horvitz, "AI in the Open World, Presidential Lecture," speech at the AAAI National Conference, Chicago, Illinois, July 2008, https://erichorvitz.com/AAAI_Presidential%20Address_Eric_Horvitz.pdf.

6. "AAAI Presidential Panel on Long-Term AI Futures," Association for the Advancement of Artificial Intelligence, 2009, https://aaai.org/about-aaai/aaai-presidential-panel-on-long-term-ai-futures-2008-2009.

7. Eric Horvitz, "One Hundred Year Study on AI: Reflections and Framing," Stanford University, 2014, https://ai100.stanford.edu/about/reflections-and-framing.

8. *The AI Index Annual Report 2024* (Palo Alto, CA: Stanford University, 2024), https://aiindex.stanford.edu/report/.

9. Partnership on AI, https://partnershiponai.org.

10. Eric Horvitz, "Reflections on AI and the Future of Human Flourishing," *AI Anthology* (2023), https://unlocked.microsoft.com/ai-anthology/eric-horvitz.

11. Eric Horvitz, *Reflections on the Status and Future of Artificial Intelligence*, Testimony Before the United States Senate, *Hearing on the Dawn of Artificial Intelligence*, Committee on Commerce Subcommittee on Space, Science, and Competitiveness (November 30, 2016), testimony: https://erichorvitz.com/Senate_Testimony_Eric_Horvitz.pdf; video: https://youtube.com/watch?v=fl-uYVnsEKc; Eric Horvitz, *AI and Cybersecurity: Rising Challenges and Promising Directions*, Hearing on AI Applications to Operations in Cyberspace before the Subcommittee on Cybersecurity of the Senate Armed Services Committee, 117th Cong., May 3, 2022, https://erichorvitz.com/Testimony_Senate_AI_Cybersecurity_Eric_Horvitz.pdf.

12. Abigail Sellen and Eric Horvitz, "The Rise of the AI Co-Pilot: Lessons for Design from Aviation and Beyond," *Communications of the Association for Computing Machinery* 67, no. 7 (June 28, 2024): 18–23, https://doi.org/10.1145/3637865.

13. Eric Horvitz, Vincent Conitzer, Sheila McIlraith, and Peter Stone, "Now, Later, and Lasting: 10 Priorities for AI Research, Policy, and Practice," *Communication of the Association for Computing Machinery* 67, no. 6 (May 7, 2024): 39–40, https://doi.org/10.1145/3637866.

14. National Academies of Sciences, Engineering, and Medicine, *Fostering Responsible Computing Research: Foundations and Practices* (National Academies Press eBooks, 2022), https://doi.org/10.17226/26507.

15. Jeannette M. Wing, "Trustworthy AI," *Communications of the ACM* 64, no. 10 (October 2021): 64–71, https://doi.org/10.1145/3448248.

CHAPTER 10

Protecting Scientific Integrity in an Age of Generative AI

Wolfgang Blau, Vinton G. Cerf, Juan Enriquez, Joseph S. Francisco,
Urs Gasser, Mary L. Gray, Mark Greaves, Barbara J. Grosz, Kathleen
Hall Jamieson, Gerald H. Haug, John L. Hennessy, Eric Horvitz,
David I. Kaiser, Alex John London, Robin Lovell-Badge, Marcia K.
McNutt, Martha Minow, Tom M. Mitchell, Susan Ness, Shobita
Parthasarathy, Saul Perlmutter, William H. Press,
Jeannette M. Wing, and Michael Witherell

Revolutionary advances in AI have brought us to a transformative moment for science. AI is accelerating scientific discoveries and analyses. At the same time, its tools and processes challenge core norms and values in the conduct of science, including accountability, transparency, replicability, and human responsibility.[1] These difficulties are particularly apparent in recent advances with generative AI. Future innovations with AI may mitigate some of these or raise new concerns and challenges.

With scientific integrity and responsibility in mind, the National Academy of Sciences, the Annenberg Public Policy Center of the University of Pennsylvania, and the Annenberg Foundation Trust at Sunnylands recently convened an interdisciplinary panel of experts with experience in academia, industry, and government to explore rising challenges posed by the use of AI in research and to chart a path forward for the scientific community. The panel included experts in behavioral and social sciences, ethics, biology, physics, chemistry, mathematics, and computer science, as well as leaders in higher education, law, governance, and science publishing and communication. Discussions were informed by commissioned papers detailing the development and current state of AI technologies; the potential effects of AI advances on equality, justice, and research ethics; emerging governance issues; and lessons that can be learned from past instances where the scientific community addressed new technologies with significant societal implications.[2]

Generative AI systems are constructed with computational procedures that learn from large bodies of human-authored and curated text, imagery, and analyses, including expansive collections of scientific literature. The systems are used to perform multiple operations, such as problem-solving, data analysis, interpretation of textual and visual content, and the generation of text, images, and other forms of data. In response to prompts and other directives, the systems can provide users with coherent text, compelling imagery, and analyses, while also possessing the capability to generate novel syntheses and ideas that push the expected boundaries of automated content creation.

Generative AI's power to interact with scientists in a natural manner, to perform unprecedented types of problem-solving, and to generate novel ideas and content poses challenges to the long-held values and integrity of scientific endeavors. These challenges make it more difficult for scientists, the larger research community, and the public to 1) understand and confirm the veracity of generated content, reviews, and analyses; 2) maintain accurate attribution of machine- versus human-authored analyses and information; 3) ensure transparency and disclosure of uses of AI in producing research results or textual analyses; 4) enable the replication of studies and analyses; and 5) identify and mitigate biases and inequities introduced by AI algorithms and training data.

Five Principles of Human Accountability and Responsibility

To protect the integrity of science in the age of generative AI, we call upon the scientific community to remain steadfast in honoring the guiding norms and values of science. We endorse recommendations from a recent National Academies report that explores ethical issues in computing research and promoting responsible practices through education and training.[3] We also reaffirm the findings of earlier work performed by the National Academies on responsible automated research workflows, which called for human review of algorithms, the need for transparency and reproducibility, and efforts to uncover and address bias.[4]

Building upon the prior studies, we urge the scientific community to focus sustained attention on five principles of human accountability and responsibility for scientific efforts that employ AI:

1. Transparent Disclosure and Attribution

Scientists should clearly disclose the use of generative AI in research, including the specific tools, algorithms, and settings employed; accurately attribute the human and AI sources of information or ideas, distinguishing between the two and acknowledging their respective contributions; and ensure that human expertise and prior literature are appropriately cited, even when machines do not provide such citations in their output.

Model creators and refiners should provide publicly accessible details about models, including the data used to train or refine them; carefully manage and publish information about models and their variants so as to provide scientists with a means of citing the use of particular models with specificity; provide long-term archives of models to enable replication studies; disclose when proper attribution of generated content cannot be provided; and pursue innovations in learning, reasoning, and information retrieval machinery aimed at providing users of those models with the ability to attribute sources and authorship of the data employed in AI-generated content.

2. Verification of AI-Generated Content and Analyses

Scientists are accountable for the accuracy of the data, imagery, and inferences that they draw from their uses of generative models. Accountability requires the use of appropriate methods to validate the accuracy and reliability of inferences made by or with the assistance of AI, along with a thorough disclosure of evidence relevant to such inferences. It includes monitoring and testing for biases in AI algorithms and output, with the goal of identifying and correcting biases that could skew research outcomes or interpretations.

Model creators should disclose limitations in the ability of systems to confirm the veracity of any data, text, or images generated by AI. When verification of the truthfulness of generated content is not possible, model output should provide clear, well-calibrated assessments of confidence. Model creators should proactively identify, report, and correct biases in AI algorithms that could skew research outcomes or interpretations.

3. Documentation of AI-Generated Data

Scientists should mark AI-generated or synthetic data, inferences, and imagery with provenance information about the role of AI in their generation, so that it is not mistaken for observations collected in the real world. Scientists should not present AI-generated content as observations collected in the real world.

Model creators should clearly identify, annotate, and maintain provenance about synthetic data used in their training procedures and monitor the issues, concerns, and behaviors arising from the reuse of computer-generated content in training future models.

4. A Focus on Ethics and Equity

Scientists and model creators should take credible steps to ensure that their uses of AI produce scientifically sound and socially beneficial results while taking appropriate steps to mitigate the risk of harm. This includes advising scientists and the public on the handling of trade-offs associated with making certain AI technologies available to the public, especially in light of potential risks stemming from inadvertent outcomes or malicious applications.

Scientists and model creators should adhere to ethical guidelines for AI use, particularly in terms of respect for clear attribution of observational versus AI-generated sources of data, intellectual property, privacy, disclosure, and consent, as well as the detection and mitigation of potential biases in the construction and use of AI systems. They should also continuously monitor other societal ramifications likely to arise as AI is further developed and deployed and update practices and rules that promote beneficial uses and mitigate the prospect of social harm.

Scientists, model creators, and policymakers should promote equity in the questions and needs that AI systems are used to address as well as equitable access to AI tools and educational opportunities. These efforts should empower a diverse community of scientific investigators to leverage AI systems effectively and to address the diverse needs of communities, including the needs of groups that are traditionally underserved or marginalized. In addition, methods for soliciting meaningful public participation in evaluating equity and fairness of AI technologies and uses should be studied and employed.

AI should not be used without careful human oversight in decisional steps of peer review processes or decisions around career advancement and funding allocations.

5. Continuous Monitoring, Oversight, and Public Engagement

Scientists, together with representatives from academia, industry, government, and civil society, should continuously monitor and evaluate the impact of AI on the scientific process, and with transparency, adapt strategies as necessary to maintain integrity. Because AI technologies are rapidly evolving, research communities must continue to examine and understand the powers, deficiencies, and influences of AI; work to anticipate and prevent harmful uses; and harness its potential to address critical societal challenges. AI scientists must at the same time work to improve the effectiveness of AI for the sciences, including addressing challenges with veracity, attribution, explanation, and transparency of training data and inference procedures. Efforts should be undertaken within and across sectors to pursue ongoing study of the status and dynamics of the use of AI in the sciences and pursue meaningful methods to solicit public participation and engagement as AI is developed, applied, and regulated. Results of this engagement and study should be broadly disseminated.

A New Strategic Council to Guide AI in Science

We call upon the scientific community to establish oversight structures capable of responding to the opportunities AI will afford science and to the unanticipated ways in which AI may undermine scientific integrity.

We propose that the National Academies of Sciences, Engineering, and Medicine establish a Strategic Council on the Responsible Use of Artificial Intelligence in Science.[5] The council should coordinate with the scientific community and provide regularly updated guidance on the appropriate uses of AI, especially during this time of rapid change. The council should study, monitor, and address the evolving uses of AI in science; new ethical and societal concerns, including equity; and emerging threats to scientific norms. The council should share its insights across disciplines and develop and refine best practices.

More broadly, the scientific community should adhere to existing guidelines and regulations while contributing to the ongoing development of public and private AI governance. Governance efforts must include engagement with the public about how AI is being used and should be used in the sciences.

With the advent of generative AI, all of us in the scientific community have a responsibility to be proactive in safeguarding the norms and values of science. That commitment—together with the five principles of human accountability and responsibility for the use of AI in science and the standing up of the council to provide ongoing guidance—will support the pursuit of trustworthy science for the benefit of all.

Notes

Originally published in the *Proceedings of the National Academy of Sciences* 121, no. 22 on May 21, 2024.

1. National Academies of Sciences, Engineering, and Medicine, *Fostering Integrity in Research* (Washington, DC: National Academies Press, 2017), https://doi.org/10.17226/21896; National Academies of Sciences, Engineering, and Medicine, *Reproducibility and Replicability in Science* (Washington, DC: National Academies Press, 2019), https://doi.org/10.17226/25303; National Academies of Sciences, Engineering, and Medicine, *Fostering Responsible Computing Research: Foundations and Practices* (Washington, DC: National Academies Press, 2022), https://doi.org/10.17226/26507.

2. Marc Aidinoff and David Kaiser, "Novel Technologies and the Choices We Make: Historical Precedents for Managing Artificial Intelligence," *Issues in Science and Technology*, 2024, https://doi.org/10.58875/BUXB2813; Urs Gasser, "Governing AI with Intelligence," *Issues in Science and Technology*, 2024, https://doi.org/10.58875/AWJG1236; Mary L. Gray, "A Human Rights Framework for AI Research Worthy of Public Trust," *Issues in Science and Technology*, 2024, https://doi.org/10.58875/ERUU8159; Alex John London, "A Justice-Led Approach to AI Innovation," *Issues in Science and Technology*, 2024, https://doi.org/10.58875/KNRZ2697; Shobita Parthasarathy and Jared Katzman, "Bringing Communities In, Achieving AI for All," *Issues in Science and Technology*, 2024, https://doi.org/10.58875/SLRG2529.

3. NASEM, *Fostering Responsible Computing Research*.

4. National Academies of Sciences, Engineering, and Medicine, *Automated Research Workflows for Accelerated Discovery: Closing the Knowledge Discovery Loop* (Washington, DC: National Academies Press, 2022), https://doi.org/10.17226/26532.

5. Patterned after the existing Strategic Council for Research Excellence, Integrity, and Trust at the NASEM, this Strategic Council will also operate in a nimble, strategic, and responsive manner to address critical issues in a fast-moving area that impacts the conduct and trustworthiness of scientific research. The narrower focus on AI will allow this second Strategic Council to focus on impacts of AI and involve users, developers, and other stakeholders in the applications of AI to scientific advancement.

CHAPTER 11

Safeguarding the Norms and Values of Science in the Age of Generative AI

Kathleen Hall Jamieson and Marcia K. McNutt

Revolutionary advances in AI have brought us to a transformative moment for science. AI is accelerating scientific discoveries and analyses. At the same time, its tools and processes challenge core norms and values in the conduct of science, including accountability, transparency, replicability, and human responsibility. . . .

We call upon the scientific community to establish oversight structures capable of responding to the opportunities AI will afford science and to the unanticipated ways in which AI may undermine scientific integrity.

We propose that the National Academies of Sciences, Engineering, and Medicine establish a Strategic Council on the Responsible Use of Artificial Intelligence in Science. The council should coordinate with the scientific community and provide regularly updated guidance on the appropriate uses of AI, especially during this time of rapid change. The council should study, monitor, and address the evolving uses of

AI in science; new ethical and societal concerns,
including equity; and emerging threats to scientific
norms. The council should share its insights across
disciplines and develop and refine best practices.
—Blau et al., "Protecting Scientific Integrity
in an Age of Generative AI"

This chapter is premised on the fact capsulized in the opening sentence of the NAS-APPC-Sunnylands (hereafter "Sunnylands" or "working group") working group statement: "Revolutionary advances in AI have brought us to a transformative moment for science."[1] Many of these transformations are explored in Chapter 3 by Jeannette Wing, Chapter 8 by Eric Horvitz and Tom Mitchell, and in the perspectives pieces in Chapter 9 (for biographies, see Appendix 2).

However, as the Sunnylands statement also notes, AI's "tools and processes challenge core norms and values in the conduct of science, including accountability, transparency, replicability, and human responsibility."[2] Here we explain the need to safeguard the interrelated scientific norms of transparency and accountability (and, with them, replicability) as well as the ethical principles that shape our understanding of scientists' responsibilities in the face of transformative changes. In the process, we signal the rationale underlying the working group's twofold call for monitoring and addressing threats to those norms and ethical values. The first urges "Scientists, together with representatives from academia, industry, government, and civil society . . . [to] continuously monitor and evaluate the impact of AI on the scientific process, and with transparency, adapt strategies as necessary to maintain integrity."[3] The second calls for establishment of a National Academy of Sciences, Engineering and Medicine (NASEM) Strategic Council on the Responsible Use of Artificial Intelligence in Science to "monitor, and address the evolving uses of AI in science; new ethical and societal concerns, including equity; and emerging threats to scientific norms."

Whether scientists are probing black holes, microbes, or human psychology, the scientific community which they form is bound together by its commitment to a common set of norms, among them one that requires individual and collective human responsibility for engaging in practices that foster accountability and with it the transparency that makes replicability possible. This combination of commitments provides scientists with the wherewithal to engage in the organized skepticism that fosters a hallmark of science, a culture of critique and correction. That culture in turn incentivizes an ongoing updating of what is known through science's methods. Scientists are united as well by ongoing efforts to ensure not only that these norms and values are honored but also that ways to honor them are refreshed in the face of changing circumstances. The rapidly evolving capacities of AI have created such circumstances.

The Nature, Function, and Importance
of Scientific Norms

Whether thought of as aspirations,[4] prescriptions telling scientists how they should behave,[5] or myths used by scientists to justify resources, enhance survivability, and burnish perceptions of the legitimacy of their work,[6] the norms espoused by science, such as accountability, transparency (and with it replicability), a culture of critique and correction,[7] and respect for the ethical limits and obligations in the conduct of their work (i.e., respect for persons, beneficence, and justice), are an integral part of scientists' self-presentation.[8] Together, they encourage scientists "to resist contrary impulses."[9] Structures that incentivize transparency and catching and correcting errors and fraud couple with the inherent competition among scientists to sustain the organized skepticism[10] that facilitates both discovery and the production and updating of knowledge.[11]

Because trust in science increases when scientists and the outlets certifying the trustworthiness of their work honor the norms and values of science,[12] the scientific community, in the form of universities, journal families, professional organizations, and entities such as the National Academy of Sciences and National Science Foundation, engages in an ongoing examination of ways to increase adherence to them. For example, the salience of the norm of transparency was bolstered when major journals began requiring preregistration of hypotheses and analysis plans, disclosure of conflicts of interest, and depositing of data and codes as conditions of publication. Along the way, signals of trustworthiness such as checklists and badging have been conventionalized as means of communicating that a publication has honored scientific norms.[13] Structures that inculcate norms and instill the value of human accountability include the responsible conduct of research (RCR) education and training that the National Institutes of Health requires of its grantees[14] and Institutional Review Boards that superintend research involving human subjects in universities.[15]

Among the norm-related themes interlaced throughout the various AI governance frameworks explored by professor of Public Policy, Governance, and Innovative Technology, and dean of the TUM School of Social Sciences and Technology at the Technical University of Munich, Urs Gasser, in Chapter 5 are those of concern here, including the need to ensure that AI systems are responsive to human needs, ethical, subject to human oversight and accountability, and transparent.

Human Accountability and Responsibility

Concerns Motivating Calls for a Focus on Human Responsibility and Accountability

There is widespread agreement that AI should "augment human intelligence, not replace it,"[16] a sentiment sometimes phrased as the desire to see AI function as a copilot not an autopilot.[17] There is agreement as well that vigorous human oversight of the development of AI is needed. Consistent with this view, a March 2023 open letter signed by CEO of Tesla Motors Elon Musk, Apple cofounder Steve Wozniak, Skype cofounder Jaan Tallinn, and a number of "well-known AI researchers"[18] called for a six-month "pause" in AI development. "AI labs and independent experts should use this pause to jointly develop and implement a set of shared safety protocols for advanced AI design and development that are rigorously audited and overseen by independent outside experts," the signatories noted.[19]

The letter asserted that "Contemporary AI systems are now becoming human-competitive at general tasks" and, in language that some saw as overblown,[20] cast dire threats to humankind in questions including "*Should* we let machines flood our information channels with propaganda and untruth? *Should* we automate away all the jobs, including the fulfilling ones? *Should* we develop nonhuman minds that might eventually outnumber, outsmart, obsolete and replace us? *Should* we risk loss of control of our civilization?" "Such decisions," it concluded, "must not be delegated to unelected tech leaders."[21]

After that letter's call for a pause went unheeded, two months later a number of the same experts issued a twenty-two-word statement that elicited headlines around the globe. "Mitigating the risk of extinction from AI should be a global priority alongside other societal-scale risks such as pandemics and nuclear war," it said. Among the goals of that statement was creating "common knowledge of the growing number of experts and public figures who also take some of advanced AI's most severe risks seriously."[22] As in the case of the letter that preceded it, the credentials of the signatories heightened the credibility of the posited risk. "Published by a San Francisco-based non-profit, the Center for AI Safety," the statement "has been co-signed by figures including Google DeepMind CEO Demis Hassabis and OpenAI CEO Sam Altman, as well as Geoffrey Hinton and Yoshua Bengio—two of the three AI researchers who won the 2018 Turing Award (sometimes referred to as the 'Nobel Prize of computing') for their work on AI," noted *The Verge*.[23]

Agreement on the Need for Human Responsibility and Accountability Expressed in Other Frameworks

The need for human responsibility and accountability has been voiced from the beginning of contemporary deliberations about the future of AI. So, for example, "Responsibility" and "Human Control" were among the Asilomar AI Principles promulgated in 2017 to guide the development of AI. In the words of that influential document:

Responsibility: Designers and builders of advanced AI systems are stakeholders in the moral implications of their use, misuse, and

actions, with a responsibility and opportunity to shape those
implications. . . .
Human Control: Humans should choose how and whether to
delegate decisions to AI systems, to accomplish human-chosen
objectives.[24]

However, efforts to ensure human responsibility and accountability are
complicated by the transformations that AI portends, the pace with which
its capacities are evolving, and the opacity of its systems. "Human agency
and oversight"[25] are as a result focal to guidelines such as the 2019 European
Union's High-Level Expert Group on AI's Ethics Guidelines for Trustworthy Artificial Intelligence. "AI systems should empower human beings, allowing them to make informed decisions and fostering their fundamental
rights," it declares. "At the same time, proper oversight mechanisms need
to be ensured, which can be achieved through human-in-the-loop, human-
on-the-loop, and human-in-command approaches."[26] The same focus can
be found in the document cast as "the world's first comprehensive AI regulatory framework,"[27] the European Union's AI Act. "As a prerequisite, AI
should be a human-centric technology," that Act states. "It should serve as
a tool for people, with the ultimate aim of increasing human well-being."[28]
In a similar manner, the November 2023 Bletchley Declaration by the
countries attending the AI Safety Summit, a list that includes the United
States, declares that "for the good of all, AI should be designed, developed, deployed, and used, in a manner that is safe, in such a way as to be
human-centric, trustworthy and responsible."[29]

Human Accountability and Responsibility in the
Sunnylands Statement

Consistent with these frameworks, the Sunnylands statement calls on relevant communities to focus sustained attention on five principles of
human accountability and responsibility for scientific efforts that use AI.[30]
The first, "transparent disclosure and attribution," focuses on transparency. The second's appeal for "verification of AI-generated content and

analysis" focuses on accountability. The third, "documentation of AI-generated data," aims to achieve both. The fourth calls for "a focus on ethics and equity." The first four define foci for the "continuous monitoring, oversight, and public engagement" called for by the fifth principle.

Transparency and Accountability

Calls for Transparency and Accountability in Other Frameworks

The context dependence of transparency and oversight requirements for AI is specifically recognized in Article 8 of the Council of Europe Framework Convention on Artificial Intelligence and Human Rights, Democracy and the Rule of Law, which notes that "Each Party shall adopt or maintain measures to ensure that adequate transparency and oversight requirements tailored to the specific contexts and risks are in place in respect of activities within the lifecycle of artificial intelligence systems, including with regard to the identification of content generated by artificial intelligence systems." Underlying the EU Act for example, are requirements directly applicable to nation states and their citizenries. These include: "AI systems and their decisions should be explained in a manner adapted to the stakeholder concerned. Humans need to be aware that they are interacting with an AI system and must be informed of the system's capabilities and limitations."[31] Likewise, a specific facet of governmental systems is the focus of the Asilomar AI principle titled "judicial transparency" that specifies that "Any involvement by an autonomous system in judicial decision-making should provide a satisfactory explanation auditable by a competent human authority."[32]

Why Transparency and Accountability Matter in Science

In science, the norm of transparency focuses on disclosure of the scientist's affiliations (e.g., verified through ORCID), contributions, and potential conflicts of interest (currently documented in contributorship and

COI attestations signed when submitting work to a journal), and on the accessibility and usability of the means needed to critique, reproduce, and replicate scientific work. By requiring transparent disclosure of an individual's contributions to the investigation as well as any relationships and interests that might bias the research process or reporting of it, science ties crediting an author to accountability for a specified facet of the work.[33] At the same time, ready access to data, methods, and code and the like make possible five activities that ground reliability in science: reproduction, replication, critique, accountability, and correction.

A study cannot be reproduced or replicated, or its errors caught and the scientific record corrected, unless its data and methods are disclosed and available for critique, reproduction, and replication by others. As the 2019 NASEM report *Reproducibility and Replicability* recognizes, "transparency [which represents the extent to which researchers provide sufficient information to enable others to reproduce the results] is a prerequisite for reproducibility."[34] To assess the validity of a study's findings, independent researchers reproduce the study. When they obtain "consistent results using the same input data; computational steps, methods, and code; and conditions of analysis," the results are considered reproducible. "To help ensure the reproducibility of computational results," NAS's 2019 *Reproducibility and Replicability in Science* report states, "researchers should convey clear, specific, and complete information about any computational methods and data products that support their published results in order to enable other researchers to repeat the analysis, unless such information is restricted by nonpublic data policies. That information should include the data, study methods, and computational environment."[35] Without reproducibility, replication in which a researcher "collects new data to arrive at the same scientific findings as a previous study" is impossible.[36]

Studies that replicate increase the body of scientific knowledge. "If findings are not replicable, then prediction and theory development are stifled," notes psychologist and founder the Center for Open Science, Brian Nosek. "If findings are replicable, then interrogation of their meaning and validity can advance knowledge. Assessing replicability can be productive for generating and testing hypotheses by actively confronting current understandings to identify weaknesses and spur innovation."[37]

In recent decades, journals have reinforced the norm of transparency and fostered critique, reproduction, and replication by requiring preregistration and public access to data, code, and analysis plans.[38] By requiring disclosure of exacting detail, they increase the ability of scholars to reproduce and replicate work. So for example, *Science* requires that authors "indicate whether there was a pre-experimental plan for data handling (such as how to deal with outliers), whether they conducted a sample size estimation to ensure a sufficient signal-to-noise ratio, whether samples were treated randomly, and whether the experimenter was blind to the conduct of the experiment."[39] And in 2024 both the *Proceedings of the National Academy of Sciences (PNAS)* and *Science* added reporting requirements pertaining to the nature of survey research samples and the ways in which they are weighted.[40]

Scientific disciplines reinforce and refine means of honoring the norm of transparency as well. In 2012, for example, the American Political Science Association (APSA) promulgated guidelines stating that researchers were ethically obligated to "facilitate the evaluation of their evidence-based knowledge claims through data access, production transparency, and analytic transparency."[41] In 2022, APSA updated those requirements to more clearly delineate the expected forms of access:

> Researchers have an ethical obligation to facilitate the evaluation of their research or empirical results. Researchers should be explicit about the data sources and methods used, including data sampling, weighting, research design, etc. Researchers should reference the data sources used. If the data were generated or collected by the scholar, researchers should provide access to those data or explain why they cannot. Researchers working with commercial data, big data, text or audio data, social media data, biometric data, digital media archives, geo-located data or confidential data sources that cannot be made publicly available in raw form should provide summary statistics of the data at the finest granulation possible, and clear coding and replication documentation. Attempts to allow others to replicate the analysis should be undertaken. Whenever possible, researchers should

provide access to the raw data. Researchers should follow scientific standards for making evidence-based knowledge claims by providing a detailed account of how they draw their analytic conclusions from the data.[42]

The Threat AI Poses to the Norm of Transparency and, with It, to Reproducibility and Replication

However, the opacity of some data-intensive AI applications makes it difficult for scholars relying on them to understand and disclose their data and decision-making processes. "Most data-intensive AI applications are essentially opaque, 'black-box' systems, and new systems capabilities are needed for users to be able to understand the decisions made by the algorithms and their potential impacts on individuals and society . . . ," noted the National Academies of Sciences, Engineering, and Medicine's 2022 report, *Fostering Responsible Computing Research: Foundations and Practices.* "Computing research has only begun to address the need for transparency of these systems."[43]

This opacity threatens the scientific norms of transparency and accountability by making it challenging to ascertain model accountability and responsibility and in the process to verify the integrity of AI-generated output including images. "I don't think I will be able to recognize a good AI-generated image anymore . . . ," noted image detection sleuth Elisabeth Bik on February in 2024, "there's probably a lot of papers being produced right now that we can no longer recognize as fake."[44] Both transparency and accountability are called into question when AI technology fabricates images[45] and data or plagiarizes.[46]

The problems associated with opacity are compounded by a second phenomenon—generative AI's ability to create convincing but fabulated "hyperrealistic content."[47] "For any complex computing system, it is hard to know whether a program does what one intends and expects it to do," noted Barbara Grosz, Higgins Research Professor of Natural Sciences, Harvard SEAS (see Chapter 9) and the lead scholar on the *Fostering Responsible Computing Research* report during the Sunnylands deliberations.

"Our current inabilities to understand why generative models produce the answers they do, and the 'hallucinations' for which they are well known, exacerbate the problem of knowing whether the code they produce actually correctly performs the functions a user intends."

A further psychological factor increases human susceptibility to the pernicious effects of hallucinations. In a phenomenon known as automation bias, humans, in the words of *Fostering,* tend to defer to "(automated) computing systems, leading to their disregarding potentially countervailing possibilities or evidence or failing to pursue them."[48] The "neutral computational certainty"[49] with which these hallucinations invent content, images, analyses, and attributions and other forms of convincing "hyperrealistic content"[50] makes computational bias difficult to counteract or blunt.

Recognizing the importance of safeguarding the transparency norm of science and with it replication, reproduction, accountability, critique and correction, the Sunnylands statement calls for: (Principle one) Transparent disclosure and attribution; (Principle two) *Verification of AI-generated content*; and, analyses, and (Principle three) *Documentation of AI-generated data.*

Ethics and Equity

Calls for Ethics and Equity in Other Frameworks

The global landscape of AI includes many AI principles initiatives, as Gasser notes (see Chapter 5). Underlying them are understandings forged from the recognition that the scientific pursuit of knowledge must never be undertaken at the expense of human dignity or autonomy. These ethical frameworks include the Rome Call for Ethics promulgated on February 28, 2020 "to promote an ethical approach to artificial intelligence,"[51] a call that grounded its commitments in the Universal Declaration of Human Rights, the milestone 1948 statement by the United Nations (General Assembly resolution 217 A) that defined "a common standard of achievements for all peoples and all nations."[52] In 2024, in a ceremony in Hiroshima's Peace Memorial Park the representatives of eleven world religions including

Buddhism, Hinduism, Zoroastrianism, Bahá'í as well as of the Abrahamic faiths added their names to the list of signatories, a list that already included representatives from tech giants such as Microsoft and IBM.[53] Signatories to the Rome Call commit to "the development of an artificial intelligence that serves every person and humanity as a whole; that respects the dignity of the human person, so that every individual can benefit from the advances of technology; and that does not have as its sole goal greater profit or the gradual replacement of people in the workplace."

The same underlying precepts can be found in the 2019 European Union's High-Level Expert Group on AI's Ethics Guidelines for Trustworthy Artificial Intelligence. In that document they take the form of a commitment to "Diversity, non-discrimination and fairness" expressed as statements that "Unfair bias must be avoided, as it could have multiple negative implications, from the marginalization of vulnerable groups, to the exacerbation of prejudice and discrimination. Fostering diversity, AI systems should be accessible to all, regardless of any disability, and involve relevant stakeholders throughout their entire life circle."[54]

In Response to Scientific Abuses, Key Ethical Principles Were Codified

Important forms of human responsibility and accountability were codified in the Nurenberg Code, the Universal Declaration of Human Rights, and the *Belmont Report* in response to abuses of science both in Nazi Germany and in the United States. With its focus on "respect for human rights, individual autonomy, and informed consent,"[55] the 1947 Nuremberg Code became "part of the infrastructure of the democratic international system that emerged after World War II."[56] In a similar vein, the 1948 Universal Declaration of Human Rights affirms "the inherent dignity and of the equal and inalienable rights of all members of the human family" and offers the reminder that "disregard and contempt for human rights have resulted in barbarous acts which have outraged the conscience of mankind."[57]

Among the science-tied abuses in the United States that gave rise to the *Belmont Report*[58] were the US Public Health Service's Tuskegee Syphilis Experiment, which scholars have characterized as a "forty year deathwatch,"[59] and governmental experiments that exposed human subjects to radiation.[60] Durable reforms shaped by the Belmont principles include the Institutional Review Boards (IRBs)[61] that govern research supported by the US government[62] and the responsible conduct of research principles (RCR)[63] that are now a taken for granted part of human subjects research in the United States.

Each of the *Belmont Report*'s basic principles—respect of persons, beneficence, and justice—led to a requirement. "Just as the principle of respect for persons finds expression in the requirements for consent, and the principle of beneficence in risk/benefit assessment, the principle of justice gives rise to moral requirements that there be fair procedures and outcomes in the selection of research subjects," noted the Report.[64] As Alex John London, K&L Gates Professor of Ethics and Computational Technologies at Carnegie Mellon University, argues in *Issues in Science and Technology*, "given the esteem the Belmont system has earned, it should be no surprise that concerned parties increasingly argue for its extension to AI innovation."[65]

Building upon the Principles of the *Belmont Report*: Ethics and Equity

In his essay in this volume, London extends the Belmont principles to include two others: nonmaleficence, "generally understood as the duty to avoid inflicting harm or imposing burdens on others," and fairness, "the duty to treat like cases alike, to apply the same rules or to follow the same process for all individuals, regardless of features or characteristics that are not directly related to some morally relevant aspect of the case."[66] The norm of fairness is reflected in the Sunnylands statement's Principle Four: *Focus on Ethics and Equity.*

London's recommendations are responsive to the need highlighted by working group member Jeannette Wing, Executive Vice President for Research and Professor of Computer Science at Columbia University, who argues in Chapter 3 that the Belmont principles of beneficence, justice, and respect for persons "need to be lifted to operate on groups of individuals, not just on individuals."[67] Among the reasons, explains

NAS-APPC-Sunnylands group member Mary Gray, Senior Principal Researcher at Microsoft Research: "even rigid conformity to Belmont principles may not ensure the interests of groups said to be represented by AI models."[68] Recognizing "the interdependence and reciprocity of human beings and the moral significance of caring for others as well as ourselves," Gray argues that "a researcher dedicated to mutuality might convene their project's multiple stakeholders, who will determine together what exactly are the risks and rewards of the research and how these will be distributed."[69] And in Chapter 7, Shobita Parthasarathy, Professor of Public Policy and Women's and Gender Studies at the University of Michigan, and Jared Katzman, PhD Student and Researcher at the University of Michigan School of Information, remind us not only of the negative consequences of the biases in AI datasets and of restriction in AI access but also that technical, organization, and legal policy AI Equity solutions exist as do ones that enhance civic capacity.

These understandings form the backdrop for the Sunnylands statement's arguments that

> Scientists and model creators should adhere to ethical guidelines for AI use, particularly in terms of respect for . . . the detection and mitigation of potential biases in the construction and use of AI systems.[70]

> *Scientists, model creators,* and *policymakers* should promote equity in the questions and needs that AI systems are used to address as well as equitable access to AI tools and educational opportunities.[71]

Group members also stressed that policymakers cannot effectively address issues of equity and justice merely by "identifying statistical biases in datasets, designing systems to be more transparent and explainable in their decision making, and exercising oversight."[72] Rather, the principle of fairness requires that AI initiatives, in the words of Parthasarathy and Katzman, grapple with "the deep-seated social inequalities that shape the landscape of technology development, use, and governance."[73]

In the Sunnylands statement, these values are particularized for the scientific community with specific calls that include:

> *Scientists* and *model creators* should take credible steps to ensure that their uses of AI produce scientifically sound and socially beneficial results while taking appropriate steps to mitigate the risk of harm. . . . *Scientists* and *model creators* should adhere to ethical guidelines for AI use, particularly in terms of respect for clear attribution of observational versus AI-generated sources of data, intellectual property, privacy, disclosure, and consent, as well as the detection and mitigation of potential biases in the construction and use of AI systems. . . . *Scientists, model creators,* and *policymakers* should promote equity in the questions and needs that AI systems are used to address as well as equitable access to AI tools and educational opportunities.[74]

The Need for Ongoing Monitoring

As the report, *Fostering Responsible Computing Research*, on which the Sunnylands statement builds, notes, "A plan for ongoing monitoring and re-evaluation by those deploying technologies or otherwise responsible for their governance is needed as research insights make their way into deployed systems and expectations and concerns shift over time."[75] Throughout the deliberations that shaped the Sunnylands statement, working group members reiterated this point. As Grosz put it, "Ethics, societal impact, and responsibility need to be addressed throughout the 'pipeline' from ideation and design to deployment."[76] Recognizing this need, the Sunnylands statement's principle five calls for ongoing monitoring, oversight and public engagement by the scientific community and disciplines within it and by a NASEM AI strategic council.

Past successes testify to the value of such ongoing vigilance by the scientific community. The original February 1975 Asilomar convening on recombinant DNA molecules is a case in point. As those drafting the convening's report noted, research involving such molecules was developing rapidly and being "applied to many different biological problems." The conferees responded with a document that not only offered principles for addressing "potential risks" but also noted that "the means for assessing and balancing risks with appropriate levels of containment will need to be reexamined from time to time." Recognizing that "it is impossible to foresee the entire range of all potential experiments and make judgments on them," they argued that "it is essential to undertake a continuing reassessment of the problems in the light of new scientific knowledge."[77] Fittingly, in 2017 the Asilomar AI Principles statement made a similar recommendation noting that "advanced AI could represent a profound change in the history of life on Earth, and should be planned for and managed with commensurate care and resources."[78] As the three International Genome Summits that are the subject of the Baltimore-Lovell-Badge case study in Chapter 2 attest, statements by the scientific community can play a role in circumscribing and guiding uses of new technologies.

Despite numerous examples of scientists and other stakeholders taking the initiative to curb abuses of new technology by articulating ethical principles, repeating past successes but now with AI could prove to be far more difficult. AI is already well developed as a commercial product by numerous large corporations, which was not the case with genome editing and other developments that presented ethical concerns. It remains to be seen whether market competition can motivate leadership and developers in commercial AI to adopt principles that guard against abuses or whether competition will disincentivize controls and guardrails.

The call for a NASEM strategic council on AI is based on the presupposition that the academies are the logical home for such an effort. Not only are these entities tasked with advising the nation on scientific matters, but they play an important role in safeguarding the norms of science. As the introduction noted, that norm protective function is evident in reports such as *On Being a Scientist: A Guide to Responsible Conduct in Research*,[79] *Fostering Integrity in Research*,[80] *Reproducibility and Replicability in Science*,[81] *Fostering Responsible Computing Research: Foundations and Practices*,[82] and workshops such as On Leading a Lab: Strengthening Scientific Leadership in Responsible Research.

Creating and superintending the strategic council on AI, as the working group urges, would also comport with the core mission of the academies. Founded by legislation passed by Congress and signed into law by President Lincoln in March 1863, NAS's mission is providing "leadership in science for the nation and the world by: Recognizing and elevating the best science and fostering its broad understanding. Producing and promoting adoption of independent, authoritative, trusted scientific advice for the benefit of society."[83]

Unsurprisingly then after reviewing the history and lessons of past efforts by the scientific community, including the recombinant DNA one, in Chapter 4, Marc Aidinoff, Research Associate at the Institute for Advanced Learning, and David Kaiser, Germeshausen Professor of the History of Science and Professor of Physics, Massachusetts Institute of Technology, draw the conclusion that forms the backdrop of the Sunnylands monitoring, oversight, and engagement principle (principle five)." Specialists and nonexpert stakeholders should regularly scrutinize both evolving technologies and the shifting social practices within which they are embedded. Only then can best practices be identified and refined."[84]

Consistent with that conclusion, governments around the globe have instituted structures to monitor and oversee the development of generative AI. The Sunnylands statement recognizes that it is both appropriate and necessary that the scientific community and the disciplines within it do the same. Protecting the integrity of science and its norms and values should be a central focus of such efforts.

Notes

Note to epigraph: Wolfgang Blau et al., "Protecting Scientific Integrity in an Age of Generative AI," *PNAS* 121, no. 22 (May 21, 2024), https://www.pnas.org/doi/10.1073/pnas .2407886121. Reprinted in Chapter 10.

1. Blau et al., "Protecting Scientific Integrity in an Age of Generative AI."

2. Blau et al., "Protecting Scientific Integrity in an Age of Generative AI."

3. Blau et al., "Protecting Scientific Integrity in an Age of Generative AI."

4. Melissa Anderson, Brian Martinson, and Raymond De Vries, "Normative Dissonance in Science: Results from a National Survey of US Scientists," *Journal of Empirical Research on Human Research Ethics* 2, no. 4 (December 2007), https://doi.org/10.1525/jer.2007.2.4.3.

5. David Hess, *Science Studies: An Advanced Introduction* (New York: New York University Press, 1997).

6. John Meyer and Brian Rowan, "Institutional Organizations: Formal Structure as Myth and Ceremony," *American Journal of Sociology* 83, no. 2 (September 1977), https://doi.org/10.1086/226550.

7. Bruce Alberts, Ralph J. Cicerone, Stephen E. Fienberg, Alexander Kamb, et al., "Self-Correction in Science at Work," *Science* 348, no. 6242 (June 26, 2015): 1420–1422, https://doi.org/10.1126/science.aab3847.

8. Yotam Ophir, Dror Walter, Patrick E. Jamieson, and Kathleen Hall Jamieson, "Factors Assessing Science's Self-Presentation Model and Their Effect on Conservatives' and Liberals' Support for Funding Science," *Proceedings of the National Academy of Sciences* 120, no. 38 (September 19, 2023), https://doi.org/10.1073/pnas.2213838120.

9. John Ziman, *Real Science: What It Is, and What It Means* (Cambridge: Cambridge University Press, 2000), https://doi.org/10.1017/CBO9780511541391.

10. Robert K. Merton, *The Sociology of Science: Theoretical and Empirical Investigations* (Chicago: University of Chicago Press, 1973).

11. Ophir et al., "Factors Assessing Science's Self-Presentation Model."

12. Arthur Lupia et al., "Trends in US Public Confidence in Science and Opportunities for Progress," *Proceedings of the National Academy of Sciences* 121, no. 11 (2024): e2319488121.

13. Kathleen Hall Jamieson, Marcia McNutt, Veronique Kiermer, and Richard Sever, "Signaling the Trustworthiness of Science," *Proceedings of the National Academy of Sciences* 116, no. 39 (2019): 19231–19236.

14. National Institutes of Health Office of Intramural Research, "Responsible Conduct of Research Training," National Institutes of Health, https://oir.nih.gov/sourcebook/ethical-conduct/responsible-conduct-research-training; "Update on the Requirement for Instruction in the Responsible Conduct of Research," National Institutes of Health, November 24, 2009, https://grants.nih.gov/grants/guide/notice-files/not-od-10-019.html.

15. "Institutional Review Boards: Actions Needed to Improve Federal Oversight and Examine Effectiveness," US Government Accountability Office, January 17, 2023, https://www.gao.gov/products/gao-23-104721.

16. David De Cremer and Garry Kasparov, "AI Should Augment Human Intelligence, Not Replace It," *Harvard Business Review*, March 18, 2021, https://hbr.org/2021/03/ai-should-augment-human-intelligence-not-replace-it.

17. Charles Chebli, "The Importance of Using AI as a Copilot, Not Autopilot: Human Intelligence Remains Essential to Artificial Intelligence!," LinkedIn, June 14, 2024, https://www.linkedin.com/pulse/importance-using-ai-copilot-autopilot-human-remains-essential-chebli-jpuzf#:~:text=The%20Role%20of%20AI%20as,verify%20AI%2Dgenerated%20insights%20independently.

18. James Vincent, "Elon Musk and Top AI Researchers Call for Pause on 'Giant AI Experiments'," *The Verge*, March 29, 2023, https://www.theverge.com/2023/3/29/23661374/elon-musk-ai-researchers-pause-research-open-letter.

19. "Pause Giant AI Experiments: An Open Letter," Future of Life, March 22, 2023, https://futureoflife.org/open-letter/pause-giant-ai-experiments/.

20. See, for example, "Expert Reaction to a Statement on the Existential Threat of AI Published on the Centre for AI Safety Website," Science Media Centre, May 30, 2023, https://

www.sciencemediacentre.org/expert-reaction-to-a-statement-on-the-existential-threat-of
-ai-published-on-the-centre-for-ai-safety-website/: "Prof Noel Sharkey, Emeritus Professor
of Artificial Intelligence and Robotics, University of Sheffield: 'AI poses many dangers to humanity but there is no existential threat or any evidence for one. The risks are mostly caused
by the Natural stupidity of believing the hype.'"

21. Vincent, "Elon Musk and Top AI Researchers."

22. "Statement on AI Risk," Center for AI Safety, n.d., https://www.safe.ai/work
/statement-on-ai-risk.

23. James Vincent, "Top AI Researchers and CEOs Warn Against 'Risk of Extinction'
in 22-Word Statement," *The Verge*, May 30, 2023, https://www.theverge.com/2023/5/30
/23742005/ai-risk-warning-22-word-statement-google-deepmind-openai.

24. "Asilomar AI Principles," Future of Life, August 11, 2017, https://futureoflife.org/open
-letter/ai-principles/.

25. High-Level Expert Group on AI (AI HLEG), *Ethics Guidelines for Trustworthy AI*,
European Commission (April 8, 2019), 2, https://www.europarl.europa.eu/cmsdata/196377
/AI%20HLEG_Ethics%20Guidelines%20for%20Trustworthy%20AI.pdf.

26. *Ethics Guidelines for Trustworthy AI*, European Commission (April 8, 2019), https://
digital-strategy.ec.europa.eu/en/library/ethics-guidelines-trustworthy-ai.

27. Shiona McCallum, Liv McMahon, and Tom Singleton, "MEPs Approve World's First
Comprehensive AI Law," BBC, March 13, 2024, https://www.bbc.com/news/technology
-68546450.

28. European Parliament, "Regulation of the European Parliament and of the Council:
Laying Down Harmonised Rules on Artificial Intelligence (Artificial Intelligence Act)
and Amending Certain Union Legislative Acts," 2024, https://artificialintelligenceact.eu
/the-act/.

29. AI Safety Summit, "The Bletchley Declaration by Countries Attending the AI Safety
Summit, 1–2 November 2023," UK Department for Science, Innovation & Technology, November 1, 2023, https://www.gov.uk/government/publications/ai-safety-summit-2023-the
-bletchley-declaration/the-bletchley-declaration-by-countries-attending-the-ai-safety
-summit-1-2-november-2023.

30. Blau et al., "Protecting Scientific Integrity in an Age of Generative AI."

31. *Ethics Guidelines for Trustworthy AI*, European Commission (April 8, 2019), https://
digital-strategy.ec.europa.eu/en/library/ethics-guidelines-trustworthy-ai.

32. "Asilomar AI Principles."

33. Marcia McNutt, "Transparency in Authors' Contributions and Responsibilities to
Promote Integrity in Scientific Publication," *Proceedings of the National Academy of Sciences*
115, no. 11 (2018): 2557–2560.

34. National Academies of Sciences, *Reproducibility and Replicability in Science* (Washington DC: National Academies Press, May 7, 2019), https://www.ncbi.nlm.nih.gov/books
/NBK547537/.

35. National Academies of Sciences, *Reproducibility and Replicability in Science*.

36. National Academies of Sciences, *Reproducibility and Replicability in Science*.

37. Brian Nosek, "Replicability, Robustness, and Reproducibility in Psychological Science," *Annual Review of Psychology* 73 (January 2022), https://doi.org/10.1146/annurev-psych
-020821-114157.

38. Edward Miguel et al., "Promoting Transparency in Social Science Research," *Science* 232, no. 6166 (January 3, 2014), https://doi.org/10.1126/science.1245317; "Nature Journals Announce Two Steps to Improve Transparency," *Nature* 555, no. 6 (February 28, 2018), https://doi.org/10.1038/d41586-018-02563-4.

39. Marcia McNutt, "Reproducibility," *Science* 343, no. 6168 (January 17, 2014), https://doi.org/10.1126/science.1250475.

40. Kathleen Hall Jamieson, Arthur Lupia, Ashley Amaya, Henry E. Brady, et al., "Protecting the Integrity of Survey Research," *PNAS Nexus* 2, no. 3 (March 2023), https://doi.org/10.1093/pnasnexus/pgad049; PNAS Author Center, "Editorial and Journal Policies," *PNAS* (n.d.), https://www.pnas.org/author-center/editorial-and-journal-policies; "*Science* Journals: Editorial Policies," *Science* (n.d.), https://www.science.org/content/page/science-journals-editorial-policies#top.

41. *A Guide to Professional Ethics in Political Science*, American Political Science Association (2012), https://www.apsanet.org/Portals/54/APSA%20Files/publications/ethicsguideweb.pdf.

42. *A Guide to Professional Ethics in Political Science*, American Political Science Association (2022), https://apsanet.org/Portals/54/diversity%20and%20inclusion%20prgms/Ethics/APSA%20Ethics%20Guide%20-%20Final%20-%20February_14_2022_Council%20Approved.pdf?ver=OshhbBcL94mq7VQiYkp9vQ%3D%3D.

43. National Academies of Sciences, Engineering, and Medicine, *Fostering Responsible Computing Research: Foundations and Practices* (Washington, DC: National Academies Press, 2022), https://doi.org/10.17226/26507.

44. Deborah Balthazar, "Q&A: The Scientific Integrity Sleuth Taking on the Widespread Problem of Research Misconduct," STAT: In the Lab, February 28, 2024, https://www-statnews-com.proxy.library.upenn.edu/2024/02/28/elisabeth-bik-scientific-integrity-research-misconduct/.

45. Jinjin Gu, Xinlei Wang, Chenang Li, Junhua Zhao, et al., "AI-Enabled Image Fraud in Scientific Publications," *Patterns* 3, no. 7 (July 8, 2022), https://doi.org/10.1016/j.patter.2022.100511.

46. Faisal Elali and Leena Rachid, "AI-Generated Research Paper Fabrication and Plagiarism in the Scientific Community," *Patterns* (March 10, 2023), https://doi.org/10.1016/j.patter.2023.100706.

47. Claire Leibowicz, "Why Watermarking AI-Generated Content Won't Guarantee Trust Online," *MIT Technology Review*, August 9, 2023, https://www.technologyreview.com/2023/08/09/1077516/watermarking-ai-trust-online/.

48. National Academies of Sciences, Engineering, and Medicine, *Fostering Responsible Computing Research: Foundations and Practices.*

49. Mike Ananny, "To Reckon with Generative AI, Make it a Public Problem," *Issues in Science and Technology* 40, no. 2 (2024): 88, https://doi.org/10.58875/EHNY5426.

50. Leibowicz, "Why Watermarking AI-Generated Content Won't Guarantee Trust Online."

51. *AI Ethics for Peace*, RenAIssance Foundation (July 10, 2024), https://www.romecall.org/ai-ethics-for-peace-hiroshima-july-10th-2024/.

52. United Nations General Assembly in Paris, *Universal Declaration of Human Rights* (December 10, 1948), https://www.un.org/en/about-us/universal-declaration-of-human-rights.

53. *AI Ethics for Peace.*

54. High-Level Expert Group on AI (AI HLEG), "Ethics Guidelines for Trustworthy AI."

55. Jonathan Moreno, Ulf Schmidt, and Steve Joffe, "The Nuremberg Code 70 Years Later," *JAMA* 318, no. 9 (2017), doi:10.1001/jama.2017.10265.

56. *Trials of War Criminals before the Nuremberg Military Tribunals Under Control Council Law, No. 10* (Washington, DC: US GPO, 1949–1953), https://science.osti.gov/-/media /ber/human-subjects/pdf/about/nuremburg_code.pdf.

57. United Nations General Assembly in Paris, *Universal Declaration of Human Rights.*

58. National Commission for the Protection of Human Subjects of Biomedical and Behavioral Research, *The Belmont Report: Ethical Principles and Guidelines for the Protection of Human Subjects of Research*, Department of Health, Education, and Welfare (April 18, 1979), https://www.hhs.gov/ohrp/regulations-and-policy/belmont-report/read-the-belmont -report/index.html.

59. James Jones, "The Tuskegee Syphilis Experiment" in *The Oxford Textbook of Clinical Research Ethics* (2008): 86–96. "Congress passed the 1974 National Research Act (Pub. L. 93–348). One result was creation of the National Commission for the Protection of Human Subjects of Biomedical and Behavioral Research. Charged with identifying "basic ethical principles that should underlie the conduct of biomedical and behavioral research involving human subjects" and with developing "guidelines which should be followed to assure that such research is conducted in accordance with those principles," the Commission generated the *Belmont Report*, whose precepts were institutionalized in the form of Institutional Review Boards."

60. National Museum of Nuclear Science and Technology, "Human Radiation Experiments," Atomic Heritage Foundation, July 11, 2017, https://ahf.nuclearmuseum.org/ahf /history/human-radiation-experiments/.

61. Kailee Kodama Muscente, "Ethics and the IRB: The History of the Belmont Report," Columbia University Teachers College, August 3, 2020, https://www.tc.columbia.edu /institutional-review-board/irb-blog/2020/the-history-of-the-belmont-report/.

62. 45 CFR 46, US Department of Health and Human Services, https://www.hhs.gov /ohrp/regulations-and-policy/regulations/45-cfr-46/index.html.

63. Logan Watts, Kelsey E. Medeiros, Tyler J. Mulhearn, Logan M. Steele, et al., "Are Ethics Training Programs Improving? A Meta-Analytic Review of Past and Present Ethics Instruction in the Sciences," *Ethics and Behavior* 27, no. 5 (2017): 351–384, https://pubmed .ncbi.nlm.nih.gov/30740008/.

64. National Commission for the Protection of Human Subjects of Biomedical and Behavioral Research, *The Belmont Report.*

65. Alex John London, "A Justice-Led Approach to AI Innovation," *Issues in Science and Technology*, May 21, 2024, https://issues.org/ai-ethics-framework-justice-london/.

66. See Chapter 6, "Challenges to Evaluating Emerging Technologies and the Need for a Justice-Led Approach to Shaping Innovation."

67. See Chapter 3, "Science in the Context of AI."

68. Mary L. Gray, "A Human Rights Framework for AI Research Worthy of Public Trust," *Issues in Science and Technology*, May 21, 2024, https://issues.org/ai-ethics-research -framework-human-rights-gray/.

69. Gray, "A Human Rights Framework."

70. Blau et al., "Protecting Scientific Integrity in an Age of Generative AI."

71. Blau et al., "Protecting Scientific Integrity in an Age of Generative AI."

72. Shobita Parthasarathy and Jared Katzman, "Bringing Communities In, Achieving AI for All," *Issues in Science and Technology*, May 21, 2024, https://issues.org/artificial-intelligence-social-equity-parthasarathy-katzman/#:~:text=To%20ensure%20that%20artificial%20intelligence,emerging%20technology%20and%20build%20it.

73. See Chapter 7, "Bringing Power In: Rethinking Equity Solutions for AI."

74. Blau et al., "Protecting Scientific Integrity in an Age of Generative AI."

75. National Academies of Sciences, Engineering, and Medicine, *Fostering Responsible Computing Research*.

76. Barbara Grosz, email message to editors, December 14, 2023.

77. Paul Berg, "Summary Statement of the Asilomar Conference on Recombinant DNA Molecules," *Proceedings of the National Academy of Science* 72, no. 6 (June 1975), https://doi.org/10.1073/pnas.72.6.1981.

78. "Asilomar AI Principles."

79. National Academy of Sciences; National Academy of Engineering; Institute of Medicine; Committee on Science, Engineering, and Public Policy, *On Being a Scientist: A Guide to Responsible Conduct in Research*, 3rd ed. (Washington, DC: National Academy of Science, 2009), https://nap.nationalacademies.org/catalog/12192/on-being-a-scientist-a-guide-to-responsible-conduct-in.

80. National Academies of Sciences, Engineering, and Medicine, *Fostering Integrity in Research* (Washington, DC: National Academies Press, 2017), https://doi.org/10.17226/21896.

81. National Academies of Sciences, *Reproducibility and Replicability in Science*.

82. National Academies of Sciences, Engineering, and Medicine, *Fostering Responsible Computing Research*.

83. "About the NAS," National Academy of Science, n.d., https://www.nasonline.org/about-the-nas/.

84. Marc Aidinoff and David Kaiser, "Novel Technologies and the Choices We Make: Historical Precedents for Managing Artificial Intelligence," *Issues in Science and Technology*, May 21, 2024, https://issues.org/ai-governance-history-aidinoff-kaiser/.

Appendix 1. List of Retreatants
Appendix 2. Biographies of Framework Authors,
Paper Authors, and Editors
Index

List of Retreatants

A Framework Addressing the Future of Artificial Intelligence in Society Pre-Meeting held virtually from November 29– 30, 2023

- Marc Aidinoff, Research Associate, Institute for Advanced Study
- Wolfgang Blau, Managing Partner, Global Climate Hub Brunswick Group
- Vinton G. Cerf, VP and Chief Internet Evangelist, Google
- George Q. Daley, Dean of the Faculty of Medicine, Harvard University
- Juan Enriquez, Managing Director, Excel Venture Management
- Urs Gasser, Professor of Public Policy, Governance, and Innovative Technology, and Dean of the TUM School of Social Sciences and Technology at the Technical University of Munich
- Susan Gonzales, Founder and CEO, AIandYou
- Mary L. Gray, Senior Principal Researcher at Microsoft Research, Faculty Associate at Harvard University's Berkman Klein Center for Internet and Society
- Mark Greaves, Executive Director, AI20250, Schmidt Futures
- Barbara J. Grosz, Higgins Research Professor of Natural Sciences, Harvard SEAS
- Kathleen Hall Jamieson, Program Director, the Annenberg Foundation Trust at Sunnylands; Director, Annenberg Public Policy Center; Elizabeth Ware Packard Professor, Annenberg School for Communication, University of Pennsylvania

- Gerald H. Haug, President Leopoldina
- John L. Hennessy, President Emeritus, Stanford University, and Chairman, Alphabet Inc.
- Eric Horvitz, Chief Scientific Officer, Microsoft
- David I. Kaiser, Germeshausen Professor of the History of Science and Professor of Physics, Massachusetts Institute of Technology
- Jared Katzman, PhD Student and Researcher at the University of Michigan School of Information
- William Kearney, Executive Director, Office of News and Public Information, National Academy of Sciences, Engineering, and Medicine
- Alex John London, K&L Gates Professor of Ethics and Computational Technologies and Director of the Center for Ethics and Policy, Carnegie Mellon University; Chief Ethicist at the Block Center for Technology and Society, Carnegie Mellon University
- Robin Lovell-Badge, Principal Group Leader and Head of the Laboratory of Stem Cell Biology and Developmental Genetics at the Francis Crick Institute
- Anne-Marie Mazza, Senior Director, National Academies of Sciences, Engineering, and Medicine
- Marcia K. McNutt, President, National Academy of Sciences
- Martha Minow, 300th Anniversary University Professor, Harvard University
- Tom M. Mitchell, Founders University Professor at Carnegie Mellon University
- Susan Ness, Former Commissioner of the Federal Communications Commission
- Shobita Parthasarathy, Professor of Public Policy and Women's and Gender Studies and Cofounder and Director of the Science, Technology, and Public Policy Program, University of Michigan
- Saul Perlmutter, Franklin W. and Karen Weber Dabby Professor, University of California Berkeley
- William H. Press, Leslie Surginer Professor of Computer Science and Integrative Biology at the University of Texas at Austin

- Harold Varmus, Lewis Thomas University Professor of Medicine, Weill Cornell Medical College
- Jeannette M. Wing, Executive Vice President for Research and Professor of Computer Science, Columbia University
- Michael Witherell, Director, Lawrence Berkeley National Laboratory

A Framework Addressing the Future of Artificial Intelligence in Society **held from February 8–10, 2024 at the Sunnylands Estate in Rancho Mirage, California**

- David Baltimore, Distinguished Professor of Biology, Caltech
- Wolfgang Blau, Managing Partner, Global Climate Hub Brunswick Group
- Vinton G. Cerf, VP and Chief Internet Evangelist, Google
- Juan Enriquez, Managing Director, Excel Venture Management
- Joseph S. Francisco, President's Distinguished Professor of Earth and Environmental Science and Professor of Chemistry, University of Pennsylvania
- Urs Gasser, Professor of Public Policy, Governance, and Innovative Technology, and Dead of the TUM School of Social Sciences and Technology at the Technical University of Munich
- Mary L. Gray, Senior Principal Researcher at Microsoft Research, Faculty Associate at Harvard University's Berkman Klein Center for Internet and Society
- Mark Greaves, Executive Director, AI20250, Schmidt Futures
- Barbara J. Grosz, Higgins Research Professor of Natural Sciences, Harvard SEAS
- Kathleen Hall Jamieson, Program Director, the Annenberg Foundation Trust at Sunnylands; Director, Annenberg Public Policy Center; Elizabeth Ware Packard Professor, Annenberg School for Communication, University of Pennsylvania
- Gerald H. Haug, President Leopoldina
- John L. Hennessy, President Emeritus, Stanford University, and Chairman, Alphabet Inc.

- Eric Horvitz, Chief Scientific Officer, Microsoft
- David I. Kaiser, Germeshausen Professor of the History of Science and Professor of Physics, Massachusetts Institute of Technology
- William Kearney, Executive Director, Office of News and Public Information, National Academy of Sciences, Engineering, and Medicine
- Alex John London, K&L Gates Professor of Ethics and Computational Technologies and Director of the Center for Ethics and Policy, Carnegie Mellon University; Chief Ethicist at the Block Center for Technology and Society, Carnegie Mellon University
- Robin Lovell-Badge, Principal Group Leader and Head of the Laboratory of Stem Cell Biology and Developmental Genetics at the Francis Crick Institute
- Anne-Marie Mazza, Senior Director, National Academies of Sciences, Engineering, and Medicine
- Marcia K. McNutt, President, National Academy of Sciences
- Martha Minow, 300th Anniversary University Professor, Harvard University
- Tom M. Mitchell, Founders University Professor at Carnegie Mellon University
- Susan Ness, Former Commissioner of the Federal Communications Commission
- Shobita Parthasarathy, Professor of Public Policy and Women's and Gender Studies and Cofounder and Director of the Science, Technology, and Public Policy Program, University of Michigan
- Saul Perlmutter, Franklin W. and Karen Weber Dabby Professor, University of California Berkeley
- William H. Press, Leslie Surginer Professor of Computer Science and Integrative Biology at the University of Texas at Austin
- Harold Varmus, Lewis Thomas University Professor of Medicine, Weill Cornell Medical College
- Jeannette M. Wing, Executive Vice President for Research and Professor of Computer Science, Columbia University
- Michael Witherell, Director, Lawrence Berkeley National Laboratory

Biographies of Framework Authors, Paper Authors, and Editors

Framework Authors

Wolfgang Blau is the managing partner of Brunswick's global climate hub and an expert in climate communications. He has previously served as president international and global chief operating officer of Condé Nast, chief digital officer and, later, president of Condé Nast International, editor-in chief of ZEIT ONLINE, and executive director of digital strategy for *The Guardian*. He currently serves as an advisor to the United Nations Climate Division, United Nations Framework Convention on Climate Change, and a trustee of Internews.org.

Vinton G. Cerf is vice president and chief internet evangelist for Google. Widely known as one of the "Fathers of the Internet," Cerf is the codesigner of the TCP/IP protocols and the architecture of the internet. He is a member of the British Royal Society, the Swedish Academy of Engineering, the Institute of Electrical and Electronics Engineers, the Association for Computing Machinery, the American Association for the Advancement of Science, the American Academy of Arts and Sciences, the International Engineering Consortium, and the National Academy of Engineering.

Juan Enriquez is the managing director of Excel Venture Management. He serves on the boards of multiple nonprofits including the National Academy of Sciences, the American Academy of Arts and Sciences, the Boston

Science Museum, Harvard Medical School, and Harvard's David Rockefeller Center. He also was the founding director of the Harvard Business School's Life Sciences Project.

Joseph S. Francisco is the President's Distinguished Professor of Earth and Environmental Science and Professor of Chemistry at the University of Pennsylvania. He is a member of the American Physical Society, the American Association for the Advancement of Science, the American Academy of Arts and Sciences, and the National Academy of Science, and the former president of both the National Organization for the Professional Advancement of Black Chemists and Chemical Engineers and the American Chemical Society.

Urs Gasser is a professor of public policy, governance, and innovative technology and the dean of the TUM School of Social Sciences and Technology at the Technical University of Munich. He serves on the board of directors of the Berkman Klein Center for Internet & Society at Harvard University as an advisor to the Organization for Economic Cooperation and Development and UNICEF.

Mary L. Gray is senior principal researcher at Microsoft Research and faculty associate at Harvard University's Berkman Klein Center for Internet and Society. She maintains a faculty position in the Luddy School of Informatics, Computing, and Engineering with affiliations in Anthropology and Gender Studies at Indiana University. She also chairs the Microsoft Research Ethics Review Program and sits on the California Governor's Council of Economic Advisors and Public Responsibility in Medicine and Research.

Mark Greaves is the executive director of AI2050. Prior to joining Schmidt Futures, Greaves was a senior leader in AI and data analytics within the National Security Directorate at Pacific Northwest National Laboratory, director of knowledge systems at Vulcan Inc., director of DARPA's Joint Logistics Technology Office, and program manager in DARPA's Information Exploitation Office. Greaves was awarded the Office of the Secretary of Defense Medal for Exceptional Public Service for his contributions to US national security while serving at DARPA.

Barbara J. Grosz is Higgins Research Professor of Natural Sciences in the Paulson School of Engineering and Applied Sciences at Harvard University. She is a member of the National Academy of Engineering and the American Philosophical Society, and a fellow of the American Academy of Arts and Sciences, the Association for the Advancement of Artificial Intelligence, the Association for Computational Linguistics, the Association for Computing Machinery, and the American Association for the Advancement of Science.

Kathleen Hall Jamieson is the Elizabeth Ware Packard Professor at the Annenberg School for Communication of the University of Pennsylvania, the Walter and Leonore Annenberg Director of the Annenberg Public Policy Center, and program director of the Annenberg Foundation Trust at Sunnylands. Jamieson is a member of the National Academy of Sciences, American Philosophical Society, a distinguished scholar of the National Communication Association, a fellow of the American Academy of Arts and Sciences, the American Association for the Advancement of Science, the American Academy of Political and Social Science, and the International Communication Association, and a past president of the American Academy of Political and Social Science.

Gerald H. Haug is the president of the German National Academy of Sciences Leopoldina and professor for climate geology at the ETH Zürich. He is a member of the Mainz Academy of Science and Literature and the Academia Europaea. He has been awarded the Gottfried Wilhelm Leibniz Prize and the Albert Maucher Prize by the German Research Foundation and the Rössler-Prize by ETH Zürich.

John L. Hennessy is the former president of Stanford University and the director of its Knight-Hennessy Scholars program. He was the inaugural Willard R. and Inez Kerr Bell Professor of Electrical Engineering and Computer Science. He is an elected member of the National Academy of

Engineering, the National Academy of Science, the American Academy of Arts and Sciences, the Royal Academy of Engineering, and the American Philosophical Society.

Eric Horvitz is Microsoft's chief scientific officer. He serves on the President's Council of Advisors on Science and Technology (PCAST) and has been an advisor to national agencies including the National Science Foundation, National Institutes of Health, Office of Naval Research, and DARPA. He is a member of the US National Academy of Engineering and the American Academy of Arts and Science, an elected fellow of the Association of Computing Machinery, Association for the Advancement of AI, and the American College of Medical Informatics. He also has served as president of the Association for the Advancement of AI (AAAI) and as a commissioner on the National Security Commission on AI (NSCAI).

David I. Kaiser is Germeshausen Professor of the History of Science and Professor of Physics at the Massachusetts Institute of Technology. His historical scholarship has been honored with the Pfizer Prize and the Davis Prize from the History of Science Society, while his physics research has received the LeRoy Apker Award from the American Physical Society, as well as election as Fellow of the APS.

Alex John London is the K&L Gates Professor of Ethics and Computational Technologies, co-lead of the K&L Gates Initiative in Ethics and Computational Technologies at Carnegie Mellon University, director of the center for ethics and policy at Carnegie Mellon University, and chief ethicist at the Block Center for Technology and Society at Carnegie Mellon University. London currently serves on the World Health Organization (WHO) Expert Group on Ethics and Governance of AI, holds an appointment to the US Health and Human Services Advisory Committee on Blood and Tissue Safety and Availability, and is a member of the US National Science Advisory Board for Biosecurity (NSABB).

Robin Lovell-Badge is a principal group leader and head of the Laboratory of Stem Cell Biology and Developmental Genetics at the Francis Crick

Institute. Lovell-Badge is an elected member of the European Molecular Biology Organization, a fellow of the Academy of Medical Sciences, the Royal Society, the Royal Society of Arts, and the Royal Society of Biology. He has been awarded the Louis Jeantet Prize for Medicine, the Armory Prize from the American Academy of Arts and Sciences, the Feldberg Foundation Prize, and the Waddington Medal of the British Society for Developmental Biology.

Marcia K. McNutt is a geophysicist and the 22nd president of the National Academy of Sciences. She has previously served as president and CEO of the Monterey Bay Aquarium Research Institute, director of the US Geological Survey and editor-in-chief of the peer-reviewed journal *Science*. She is a fellow of the American Geophysical Union, Geological Society of America, the American Association for the Advancement of Science, and the International Association of Geodesy, and a member of the National Academy of Sciences, National Academy of Engineering, the American Philosophical Society, the American Academy of Arts and Sciences. Internationally, she is a foreign member of the Royal Society, the Russian Academy of Sciences, the Chinese Academy of Sciences, and the Indian National Science Academy.

Martha Minow is the 300th Anniversary University Professor at Harvard University and former dean of Harvard Law School, where she has taught since 1981. She is among two dozen individuals who bear university professorship, Harvard's highest academic post, authorizing her to pursue research and teaching at any of Harvard's schools. She is a fellow of the American Academy of Arts and Sciences, the American Bar Foundation, and the American Philosophical Society.

Tom M. Mitchell is the Founders University Professor at Carnegie Mellon University, where he founded the world's first Machine Learning Department. He co-chairs a US National Academies study on AI and the future of work, as well as a task force studying Generative AI for the Special Competitive Studies Project. Mitchell also is an elected member of the US National Academy of Engineering, and the American Academy of Arts and

Sciences, and a fellow and past-president of the Association for the Advancement of Artificial Intelligence (AAAI).

Susan Ness is a former member of the US Federal Communications Commission, where she played a leading role on spectrum policy, international ICT advocacy, competition policy, and emerging technologies. She founded Susan Ness Strategies, a digital tech and media policy consulting firm, and is a frequent convener and speaker on tech governance issues. She is also a distinguished fellow at the Annenberg Public Policy Center of the University of Pennsylvania, a nonresident senior fellow at the Atlantic Council, and a member of the World Economic Forum's Global Coalition for Digital Safety.

Shobita Parthasarathy is professor of public policy and women's and gender studies, and cofounder and director of the science, technology, and public policy program at University of Michigan. Her book *Building Genetic Medicine: Breast Cancer, Technology, and the Comparative Politics of Health Care* (MIT Press, 2007) influenced the 2013 US Supreme Court case that determined human genes were not patentable. She has held fellowships from the American Council for Learned Societies, the Woodrow Wilson International Center for Scholars, American Bar Foundation, and Max Planck Institute for Innovation and Competition (Germany).

Saul Perlmutter is a professor of physics at UC Berkeley, where he holds the Franklin W. and Karen Weber Dabby Chair, and a senior scientist at Lawrence Berkeley National Laboratory. He is the leader of the international Supernova Cosmology Project, director of the Berkeley Institute for Data Science and executive director of the Berkeley Center for Cosmological Physics. He is a member of the National Academy of Sciences, the American Philosophical Society, and the American Academy of Arts and Sciences, and a fellow of the American Physical Society and the American Association for the Advancement of Science. He currently serves on the President's Council of Advisors on Science and Technology. In 2011, he was awarded the Nobel prize in physics, sharing the prize for the discovery of the accelerating expansion of the universe.

William H. Press is the Leslie Surginer Professor of Computer Science and Integrative Biology at the University of Texas at Austin. At UT, his affiliations include membership in the Oden Institute for Computational Engineering and Sciences and in the Institute for Cellular and Molecular Biology. Press is also a senior fellow (emeritus) at the Los Alamos National Laboratory, a former member of President Barack Obama's Council of Advisors on Science and Technology (PCAST), and the past (2012–2013) president of the American Association for the Advancement of Science. He is currently the elected treasurer of the National Academy of Sciences and is a member of the Governing Board of the National Research Council.

Jeannette M. Wing is the executive vice president for research and professor of computer science at Columbia University. She is a member of the American Academy for Arts and Sciences Board of Directors and Council; the New York State Commission on Artificial Intelligence, Robotics, and Automation; and the Advisory Board for the Association for Women in Mathematics. She is also a member of the National Academy of Engineering and a fellow of the American Academy of Arts and Sciences, American Association for the Advancement of Science, Association for Computing Machinery (ACM), Institute of Electrical and Electronic Engineers (IEEE), and National Academy of Innovators. She holds an honorary Doctor of Technology from Linkoping University, Sweden.

Michael Witherell is director of the Lawrence Berkeley National Laboratory (Berkeley Lab). He previously served as vice chancellor for research for the University of California, Santa Barbara (2005–2014) and director of Fermi National Accelerator Laboratory (1999–2005). He is the recipient of the American Physical Society's W. K. H. Panofsky Prize in Experimental Particle Physics. He is a member of both the National Academy of Sciences and the American Academy of Arts and Sciences and has served on numerous boards, including as a member of the National Academies of Science, Engineering, and Medicine's Committee on Science, Engineering, Medicine, and Public Policy (2016–2019).

Paper Authors

For those who were also framework authors, see biographies above.

Marc Aidinoff is a researcher at the Institute for Advanced Study. He recently served as chief of staff and senior advisor in the White House Office of Science and Technology Policy where he helped lead a team of 150 policymakers on key initiatives including the Blueprint for an AI Bill of Rights and guidance to ensure federally funded research is publicly accessible.

David Baltimore is President Emeritus and Distinguished Professor of Biology at Caltech. He is a member of the National Academy of Sciences, fellow of the American Academy of Arts and Sciences, and a foreign member of the Royal Society of London and the French Academy of Sciences. He has also been president and chair of the American Association of the Advancement of Science. In 1975, he was awarded the Nobel Prize in Physiology or Medicine for his research in virology.

Jared Katzman is a Ph.D. student at the University of Michigan's School of Information with a specialization in Science, Technology, and Public Policy (STPP). Recently, Katzman has worked with the Center for Democracy and Technology as a research policy fellow, where they investigated gaps in the current ecosystem of AI regulation.

Editors

Kathleen Hall Jamieson (See biography above)

Anne-Marie Mazza is senior director of the National Academies' Committee on Science, Technology, and Law. From 2021 to 2022, she was detailed to the White House as executive director, President's Council of Advisors on Science and Technology. Mazza joined the National Academies of Sciences, Engineering, and Medicine in 1995 and has been the study director on numerous National Academies' activities involving emerging technologies (e.g., human genome editing, synthetic biology, neural organoids and chimeras); science in the courtroom (e.g., eyewitness identification and forensic science); and governance of academic research (e.g., dual-use research of concern, intellectual property, and human subjects). She is a fellow of the American Association for the Advancement of Science.

William Kearney is executive director of the Office of News and Public Information at the National Academies of Sciences, Engineering, and Medicine, where he also is editor of *Issues in Science and Technology*, which is copublished with Arizona State University. Kearney has led media outreach for hundreds of science, technology, and health policy reports, and managed communications for large scientific conferences including the famous international summits on human genome editing in Washington in 2015 and in Hong Kong in 2018, as well as Nobel Prize Summits in 2021 and 2023. He directed other international science communication efforts, including for the African Science Academy Development Initiative, and he has presented at the World Conference of Science Journalists. In 2010, he was assigned to the InterAcademy Council to manage communications for a review of the Intergovernmental Panel on Climate Change.

INDEX

AAAI (Association for the Advancement of AI), 5, 147, 206

AI (artificial intelligence), definition and terminology, 5, 59–60, 85; future trends, 127–128, 138, 177, 188, 208–209; historical, 148, 211; limitations, 2, 50, 71, 168, 188

AI and facial recognition, 45, 48–49, 71, 127–129, 131–132, 134, 137

AI and society, 112, 119, 123, 204, 210

AI Bill of Rights, 8, 61, 127

AI governance, 49–50, 57–59, 69, 72–75, 84, 139; challenges and future directions, 60, 63, 67, 88–89; ethical guidelines and principles, 88, 92, 178; global governance frameworks, 63–64, 70–71, 79; norms and patterns in governance, 58, 62, 68, 72, 74, 76–78, 82–84, 88–89, 92, 105–106, 113; social imperatives, 102–104, 114–115, 119–120, 122

AI in climate and sustainability, 31, 67, 138, 183–184, 188, 212, 216–217

AI in computer science, 21–22, 30, 135, 148, 178, 196, 201–202, 214–216; in computer science engineering, 203–206

AI in education, 49, 68, 103, 113, 168, 179, 184, 186, 200

AI in engineering, 170, 184

AI in healthcare, 180, 182, 183

AI in public policy, 3, 72, 76, 81, 84, 92, 109–110, 115

AI in science, 179–181, 183, 195–196, 222; astronomy, 29–30; chemistry, 199–201; geology, 211–212; physical sciences, 183, 185; physics, 31, 212–213, 216–218

AlphaFold, 1, 24, 179, 196, 203

Amazon, 133, 178

Annenberg Foundation Trust at Sunnylands, 2–4, 147, 221, 230, 242–243

Apple, 133, 176, 232

Asilomar Conference on Recombinant DNA: in 1975, 3, 36, 42–43, 243–244; in 2009, 206–207; in 2017, 233–34, 243

Belmont Report, 3, 44, 107, 216, 240–242

Biotechnology, 36, 41, 43

Bletchley Declaration, 64, 78–80, 234

Brazil's Draft Artificial Intelligence Act, 61, 70–73, 77, 79, 81–83, 86

Canada's Draft Artificial Intelligence and Data Act, 60, 70–72, 77, 82

ChatGPT. *See* OpenAI

China's Interim Generative AI Measures, 60–61, 70, 75

CoE (Council on Europe), 57, 64, 78

CRISPR-Cas9, 3, 15–17. *See also* Human genome editing

DALL-E/DALL-E2, 152, 184, 187

DNN (deep neural networks), 22–25, 27–29, 153–157, 167, 181

Equity in AI, 116–118, 123, 128, 133–134, 138–139, 224, 239, 241; biases, 48–49, 112, 117–118, 130–131, 134; digital divide, 187; equity and inclusion frameworks, 129, 134–136; funding, 132–133, 138–139, 178; policy recommendations, 35–36, 40, 64, 87

Ethics, 105–109, 110–111, 115; in AI, 108, 117, 121, 136, 224, 239, 241; ethical frameworks and principles, 64, 79, 101; ethical Guidelines, 101–102, 105, 122; malevolent uses of AI, 184, 186–187, 210, 212; trustworthy practices, 99

EU AI Act, 57, 59, 61, 63, 66–68, 70–74, 77, 80–84, 86, 91, 234–235

Facebook, 131–133

FBI, 38, 40, 131

Frameworks, AI, 65, 105–107, 119, 123, 174, 232

GDPR (General Data Protection Regulation), 63, 68, 75, 85, 91

Generative AI, 1, 19, 21, 23, 112, 158, 197, 222

Google, 129–131, 133–134, 176, 178, 197; DeepMind, 233; Gemini (formerly Bard), 162, 170, 175–176, 178

G7 and G20, 57, 64, 66, 70, 78, 83, 87

Human genome editing, 4–5, 15–19, 196, 244

Human responsibility in AI, 100, 130–131, 138–139, 206, 222–226, 232–233, 236–237

Informed consent, 4, 44, 109, 129–130, 240

IRB (Institutional Review Board), 3–4, 44, 108–110, 241

Justice in AI, 8, 64–65, 78, 101–109, 112–118, 120–123, 127, 139, 222, 231–232, 241–242

LLM (large language models), 23, 25, 29–30, 130, 168, 171–172, 175, 183; training, 198–199

Machine learning, 22; diffusion modeling, 23, 25, 29, 152, 180; discriminative and generative models, 151–152; foundations and advancements, 150–153; open-source modeling, 177, 179; transformers, 25, 27–28, 30, 152, 158–62, 164–165, 167–168, 179

Microsoft, 5, 130, 133, 174, 176, 178, 183, 240, 242

Monitoring and oversight, 12, 36, 68, 71, 86, 110, 119, 128, 130, 208, 224–226, 229–230, 235

NAS (National Academy of Sciences), 2–4, 147, 195–199, 209, 221, 226, 230–231, 236, 238–239, 243–245

NIH (National Institutes of Health), 3, 42–44, 110, 232, 238–239

NIST (National Institute of Standards and Technology), 48, 61, 66, 69, 83–84

NSF (National Science Foundation), 31, 110, 132, 179, 231

OECD (Organisation for Economic Co-operation and Development), 59–61, 64, 66, 70, 81, 85, 87

OpenAI, 1, 24–25, 27, 159, 162, 173, 178, 197, 233; GPT-4, 23, 163–171, 175–176, 182, 184–186, 207; GPT-4V, 170–171

Policymakers, 2, 8, 35, 39–43, 66, 111–113, 120–121, 127, 133–134, 222–225, 242

Rome Call for Ethics, 239–240

Scientific community, 2–5, 30, 49, 204; integrity, 1, 4, 12, 19, 36, 42, 50, 221; scientific norms, 4, 13, 19, 36–38, 57, 221, 230–231, 243

Singapore's Model AI Governance Framework, 78, 80, 83, 85

Transparency in AI, 2, 9, 64, 68, 73, 79, 127, 138, 222–225, 232–237

Turing Test, 21, 28, 148

US Executive Order on Safe, Secure, and Trustworthy AI, 57, 61, 68, 72, 77–82, 85–86, 90–91, 139

Verification of AI-generated content, 12, 168, 203, 223–224, 234–235

Printed in the USA
CPSIA information can be obtained
at www.ICGtesting.com
JSHW011047081124
73177JS00010B/252